UNIVERSE
IN FOCUS

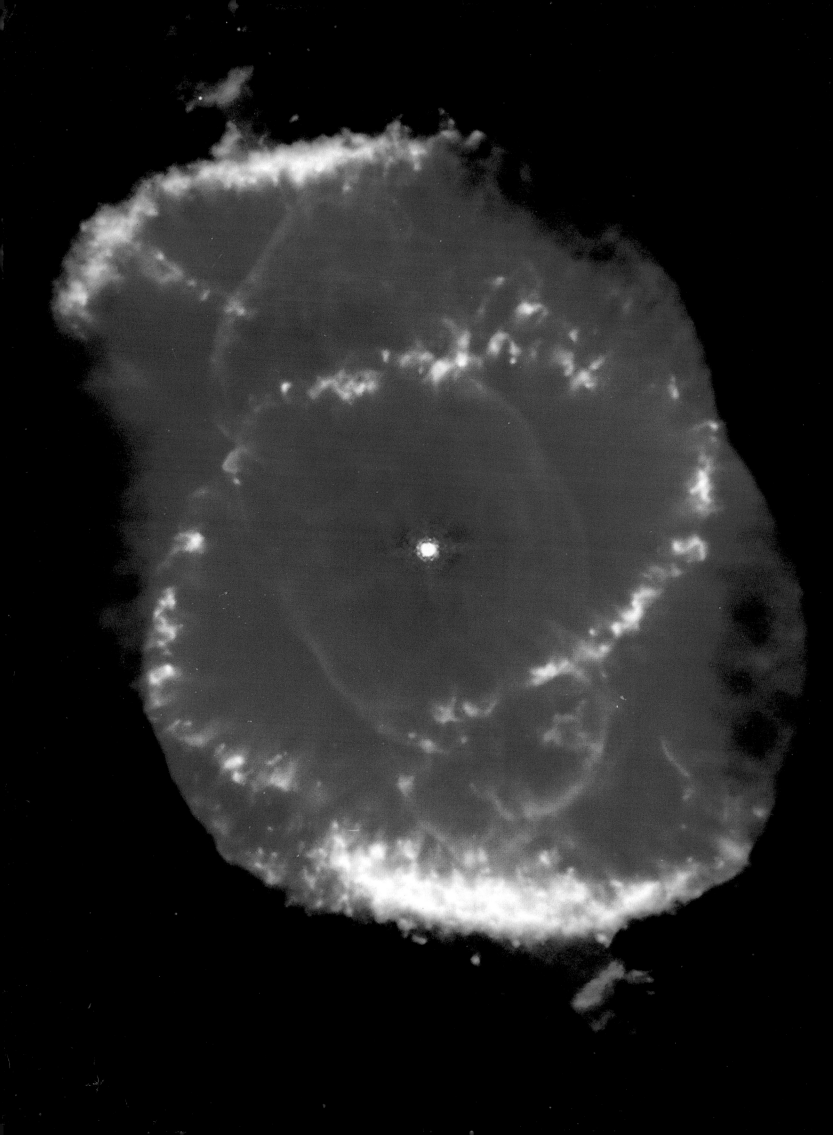

THE STORY OF THE HUBBLE TELESCOPE

UNIVERSE IN FOCUS

STUART CLARK

CASSELL

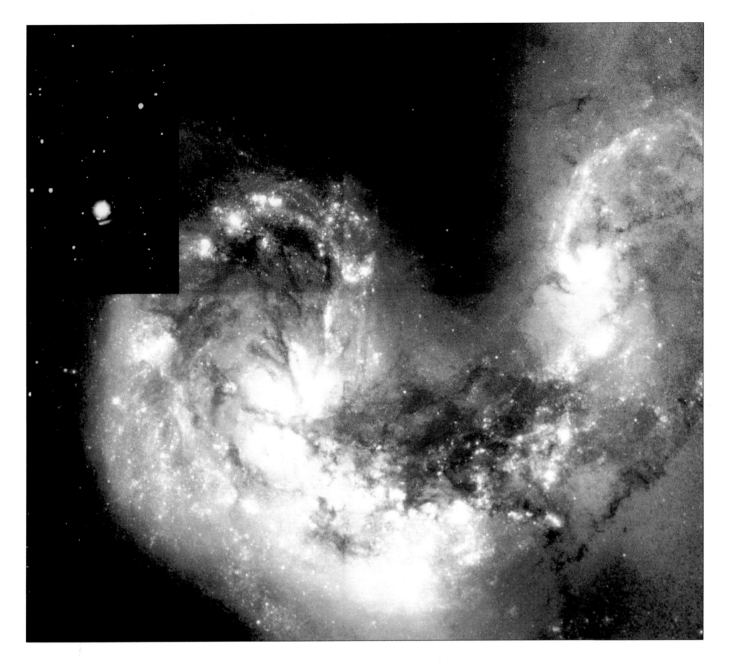

Project Editor   Fiona Gold
Art Editor & Designer   Frankie Wood
Advisory Editor   Robin Kerrod
Proofreader & Indexer   Helen McCurdy
Editorial Assistant   Marian Dreier
Picture Research Manager Claire Turner
Production   Clive Sparling
Publishing Director   Graham Bateman

 Planned and produced by
Andromeda Oxford Ltd.
11–15 The Vineyard, Abingdon
Oxfordshire
England OX14 3PX

AN ANDROMEDA BOOK

Origination by Pixel Tech Prepress Ltd., Singapore
Printed by K.H.L Printing Ltd., Singapore

FIRST PUBLISHED IN THE UK BY CASSELL IN 1997
Wellington House, 125 Strand, London WC2R OBB

This revised edition published 1998

ISBN 0–304–35025–7

British Library Cataloguing in Publication Data
A catalogue record for this book is available from the British Library

INTRODUCTORY PICTURES
Title page: The Cat's eye nebula. Ejected gas has been drawn
out into filaments and tendrils that surround the central star
producing a catherine wheel effect. Astonomers believe this
effect is caused by a, as yet undetected, companion star.

This page: A galaxy collision between two galaxies known as the
Antennae. The two large yellow blobs are the centers of the
respective galaxies, and the dark swirls are dust lanes.

CONTENTS

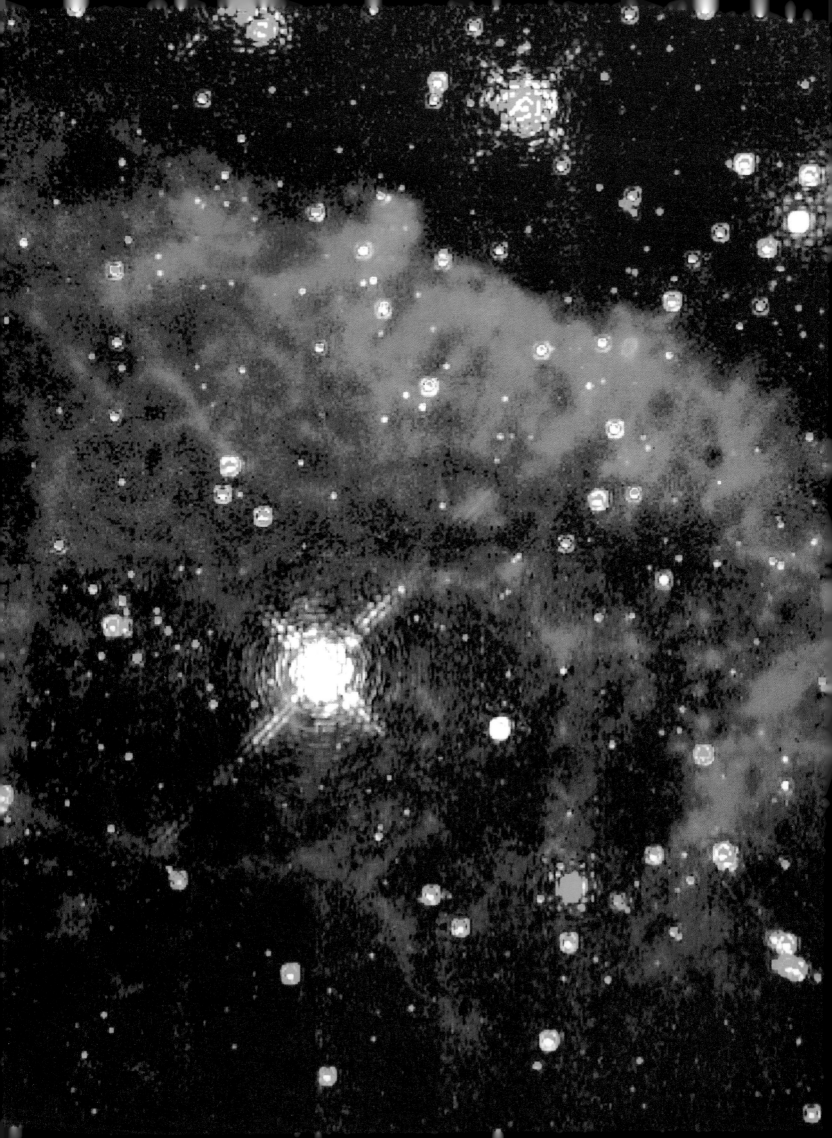

Preface
by Stuart Clark

Throughout the centuries, every practising astronomer has been tempted to believe that he or she is living in a "golden age" of the subject. Usually the feeling has been created by the intellectual stimulation surrounding the most important discoveries of the era. History has made me wary of making that claim, but I believe that astronomy is currently in a boom period. One of the major contributors to that boom is, undoubtedly, the Hubble Space Telescope. Despite the difficulties encountered soon after its launch, the repaired Space Telescope has enabled a quantum leap in observational astronomy. Launched in 1990, Hubble has taken over 100,000 separate images of the cosmos for astronomers to collate and comb for information. Many shown in this book are composites from multiple exposures. Some show familiar objects in more detail than ever before; others break new ground, probing the mysteries of black holes, quasars and the origin of the galaxies. A number of images revealed objects that, until now, scientists could only imagine.

Undoubtedly, the starting point for this book was the images themselves. All were extraordinary; some were also stunningly beautiful. The story of cosmic evolution is as fascinating as the pictures, and so I wanted to write a commentary to the images that was also a basic, yet complete, overview of the subject of modern astronomy. Because the Space Telescope is such a versatile observing tool, this whole story can be told through Hubble's own images. Occasionally, very complex images are explained in depth using attractive artwork. Only in a few places, and then only for comparison, do we see images from ground-based telescopes or other cameras. In essence, this book proves that their are no fundamental aspects of modern astronomy that Hubble has not explored and enhanced.

Universe in Focus has been organized into 6 chapters, beginning close to the Earth and working out into deep space. Chapter one focuses on the history of the Space Telescope, and the spectacular repair mission to refocus its faulty optics. Hubble's view of the other planets in our Solar System is explored in chapter two, before shifting the emphasis to the evolution of stars in chapter three. In a step up the distance ladder, the galaxies are analyzed in chapter four, before the universe, as an evolving landscape, is considered in chapter five. This section contains some of the most amazing and unprecedented visual records of our modern astronomical age. The final chapter, the Factfile, provides key reference information to expand and supplement the main content of the book.

The human imagination and endeavor that placed the Space Telescope in orbit is worthy of a book of its own. When a defect was discovered in its primary mirror, the skillful dedication of the individuals who corrected the fault is among the most remarkable achievements of modern science. They triumphed beyond the most optimistic expectations to produce a piece of technology that will be remembered forever in the annals of science.

The Pistol nebula *(left)* surrounds one of the most massive stars yet discovered at 100 times the mass of our Sun. The magenta glow is hydrogen gas which has exploded from the star during two eruptions which date from 6,000 and 4,000 years ago.

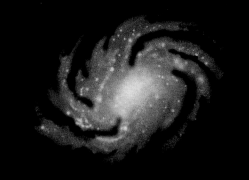

HIERARCHY OF THE UNIVERSE

THE UNIVERSE IS A VAST PLACE, overwhelming in its sheer size. In their study of the night sky, astronomers have identified many different types of celestial objects, including planets, stars and galaxies. These can be ranked in a natural hierarchy of size. On the smallest celestial scale are the planets, found in orbit around stars. Our own Solar System is made up of nine planets in orbit around a central star; the Sun. Stars are found in conglomerations known as galaxies, of which our Milky Way is just one example. Galaxies are so large that their forces of gravity cause them to clump together into groups and clusters. These clusters, in turn, group together to form superclusters, the largest known structures in the universe.

A spiral arm *(below)* of our galaxy, the Milky Way, which contains stars a few thousand light years away.

Stars in the night sky vary widely in size and color and are, typically, a few hundred light years away.

Globular Clusters are spherical groups of old stars surrounding the main stellar content of our galaxy. Similar groups are found in other galaxies.

Saturn's orbit

Pluto's orbit

Hubble's orbit

Earth

Planets in orbit around the Sun. Our Solar System contains nine planets, including Earth.

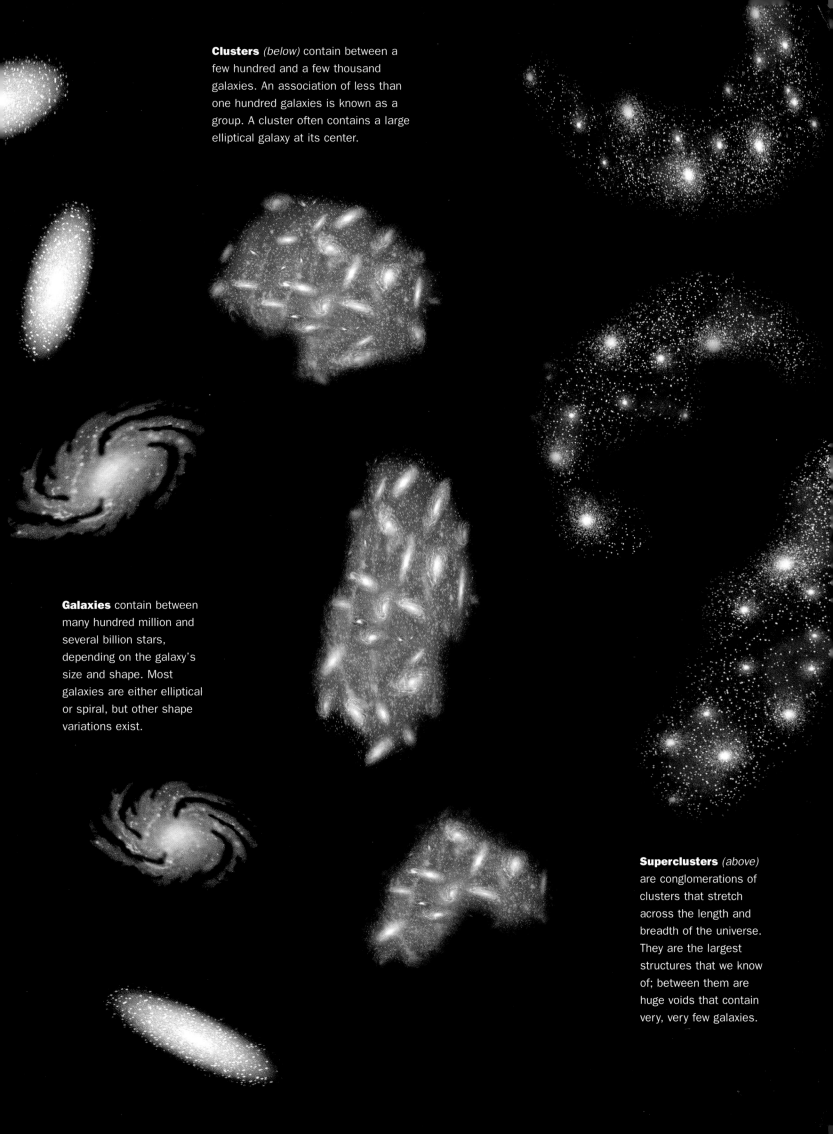

Clusters *(below)* contain between a few hundred and a few thousand galaxies. An association of less than one hundred galaxies is known as a group. A cluster often contains a large elliptical galaxy at its center.

Galaxies contain between many hundred million and several billion stars, depending on the galaxy's size and shape. Most galaxies are either elliptical or spiral, but other shape variations exist.

Superclusters *(above)* are conglomerations of clusters that stretch across the length and breadth of the universe. They are the largest structures that we know of; between them are huge voids that contain very, very few galaxies.

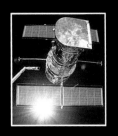

THE
HUBBLE
SPACE
TELESCOPE

*T*HE *HUBBLE SPACE TELESCOPE* is one of the finest technological achievements of the 20th century. Although not the largest telescope ever built, Hubble is uniquely located above the distorting effects of our world's atmosphere. From this vantage point it has a remarkably clear view across the universe, imaging many of the most interesting objects in space. It has captured the public's imagination with its unprecedented pictures and with the extraordinary repair mission that was required to correct its faulty optics.

FROM CONCEPT TO LAUNCH

Edwin Hubble *(above)* sits at the eyepiece of the 100 inch (2.54 meter) Hooker telescope at Mount Wilson, California. The primary mirror of the Hubble Space Telescope and the primary mirror of the Hooker telescope are comparable in size but, because of its vantage point, the Hubble Space Telescope far out-performs its ground-based counterpart.

The Space Shuttle Discovery lifts off *(right)* in April 1990, carrying the Hubble Space Telescope into orbit. Shortly after the Shuttle reached its designated altitude, it opened the cargo bay doors and deployed the telescope.

THE HUBBLE SPACE TELESCOPE is named after the distinguished American astronomer, Edwin Hubble (1889–1953). In the 1920s he carried out groundbreaking research, proving that galaxies were not clouds of gas but separate and distinct collections of stars. He also classified galaxies according to shape (see page 78). The crowning achievement in his research was the discovery that most galaxies seem to be moving away from us. Only the nearest are bound, by gravity, to the Milky Way. This result was interpreted, according to Einstein's theory of relativity, to mean that the universe is expanding.

Around the same time, in 1923, a book was published in Germany called *Die Rakete zu den Planetenraümen* (The Rocket into Planetary Space). In it the author, Hermann Oberth, described a space station with a telescope. He developed his idea by proposing that the telescope be placed on an asteroid, so that it would have a steady base from which to take images. The first serious attempt to justify a space telescope was made in 1946 by astronomer Lyman Spitzer. He submitted a proposal to the Douglas Aircraft Company for an extraterrestrial observatory. His report showed remarkable insight and is now viewed, retrospectively, as the birth of the idea for the Hubble Space Telescope. NASA did not consider in earnest the prospect of building a space-based telescope until the 1970s. As a preliminary step they had launched two small space observatories, OAO-II and Copernicus. When these were successful, serious research was invested in a large telescope

designed to perform for many years. Throughout the decade plans were drawn and redrawn. Eventually it was agreed that the cost-effective size of the primary mirror would be 95 inches (2.4 meters). The specifications for the cameras and other detectors were finalized and, in 1979, construction began. The telescope was scheduled to be launched from a Space Shuttle in 1986, but as the launch date drew near, disaster struck. The Space Shuttle Challenger exploded shortly after lift-off on January 28, 1986 and NASA halted Shuttle flights, pending an investigation. Finally, on April 24, 1990, the Space Shuttle Discovery lifted off from its launch pad at the Kennedy Space Center. It carried, in its cargo bay, the Hubble Space Telescope, our "window on the universe".

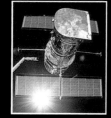

How Hubble Works

HUBBLE'S DESIGN IS SIMILAR to most large ground-based telescopes. It focuses by reflecting light off two mirrors. The larger, primary mirror focuses light onto a smaller secondary mirror near the top of the telescope. The secondary mirror reflects light back along the telescope and through a hole in the center of the primary mirror. Focused light passes into the instrument cluster, where it is routed to a selected instrument. Hubble is designed to carry five instruments, two of which are cameras. The Faint Object Camera (FOC) was designed to look in great detail at small areas, while the Wide Field and Planetary Camera (WF/PC) has a wider field of view, to capture larger objects. In addition to the cameras, NASA incorporated two spectrographs to investigate the chemical composition of celestial objects. Originally they were the Faint Object Spectrograph (FOS) and the Goddard High Resolution Spectrograph (GHRS); subsequently they were replaced by the Near-Infrared Camera and Multi-Object Spectrometer (NICMOS) and the Space Telescope Imaging Spectrograph (STIS). The fifth instrument is the High Speed Photometer (HSP), measuring variation in brightness of a celestial object.

Hubble is in a low Earth orbit, at an altitude of only 320 miles (515 km). This orbit places the Space Telescope within easy reach of NASA's Space Shuttles for servicing and updating with new cameras and instrumentation. The drawbacks of a low Earth orbit are that Hubble loses height over time, and needs its orbit to be boosted during servicing. Also, Hubble cannot stay in direct contact with its ground-based facilities as it is in direct range for only about 20 minutes during each 90-minute orbit of the Earth. In order to maintain constant communication, it uses a rather convoluted system. The telescope relays its data through two orbiting satellites from the Tracking and Data Relay Satellite System (TDRSS). Both satellites are in a much higher orbit (22,000 miles or 35,400 km), known as geostationary orbit, which means that each remains above the same location on Earth all the time. They send the information to their base at White Sands, New Mexico. From there, the data is relayed by a communications satellite to the Goddard Space Flight Center near Washington, D.C. They process the flight and engineering telemetry from the telescope and the raw scientific data is relayed by telephone link to the Space Telescope Science Institute, Baltimore.

The Hubble Space Telescope *(right)* in the payload bay of the Space Shuttle Endeavour during the repair mission. The side-on view *(top left)* shows the aperture door that shields cameras and instrumentation from the Sun. When the Sun is behind Hubble, the door lifts to allow images to be taken.

1 Primary Mirror

2 Secondary Mirror

3 Solar Panels for power.

4 Communications Antenna

5 Wide Field and Planetary Camera

6 Faint Object Camera

7 Near-Infrared Camera and Multi-Object Spectrometer (originally the Faint Object Spectrograph).

8 High Speed Photometer

9 Space Telescope Imaging Spectrograph (originally the Goddard High Resolution Spectrograph).

10 Onboard Computer for processing flight data.

11 Science Computer for processing scientific data.

12 Data Management Unit for controlling data within the telescope.

THE REPAIR MISSION

Underwater practise *(above)*. During extensive training to perfect the techniques used during the repair mission, giant swimming pools were used to simulate a weightless environment. In this underwater shot, astronauts Story Musgrave and Jeffrey Hoffman manipulate an accurate model of the Wide Field and Planetary Camera, practising how to slot it into place on a mock-up. Conventional divers can be seen in the background monitoring the training exercise.

THE LAUNCH OF THE HUBBLE SPACE TELESCOPE was a major event in the history of NASA; it opened up a new frontier of astronomy. On May 20, 1990, Hubble experienced "first light", an almost hallowed event for telescope engineers. It denotes the first time a telescope is fully operational and takes an image. Hubble's first-light ceremony was played out before hoards of media reporters. As Hubble focused on the universe, the eyes of the world turned to the telescope itself. Star cluster NGC 3532 was chosen to be the first subject; it was to be imaged by the Wide Field and Planetary Camera. In front of anxious scientists and excited reporters, the first-light image of stars in NGC 3532 flashed onto the screen. Scientifically and aesthetically, it was not an awe-inspiring picture but the very fact that the Hubble Space Telescope was working made it remarkable.

Those watching the screens were elated but, behind the scenes, the people who had struggled to produce the image were worried. The very next day, a meeting was held at which first suspicions were voiced that the mirror was the wrong shape. A month of intensive investigations followed, which revealed that the primary mirror had indeed been ground into the wrong shape. Unfortunately, the press gave Hubble more coverage over this bad news, than when they assumed that the telescope had been working correctly. As a result, the Hubble Space Telescope entered the public consciousness as a conspicuous failure; "a $1.5 billion dollar blunder" as one newspaper headline read.

The problem was a spherical aberration in the primary mirror. The rays of light reflecting from the central region were focused in a different position from those striking the outer regions of the mirror. In short, although the image should have been concentrated into a single, central point, the spherical aberration was smearing it over a much larger area. While the investigation to discover why

the mirror had been ground incorrectly continued, scientists began working on ways to fix the faulty optics. Serious consideration was given to bringing Hubble back down from orbit to Earth. This way, the repairs could be made in the controlled environment of a laboratory. If this were to happen, however, NASA would have to relaunch the Space Telescope, using a third precious Shuttle mission. Far more palatable was the option of a repair in orbit. It would require some of the longest space walks ever made but, if successful, would be a milestone in NASA's quest for people to live and work in space. In parallel to this discussion was the nature of the repair. Eventually, it was decided that a new instrument could be built that would refocus the light before it entered the instrument cluster at the base of the telescope. One of those instruments would have to be sacrificed in order for there to be room for the corrective optics unit, but that would be a small price to pay for the

The real thing *(right)*. Astronaut Kathryn Thornton has no problem lifting the bulky repair unit in the weightlessness of space. She is attached to the Space Shuttle by foot restraints on Endeavour's robotic arm. The unit contains the Corrective Optics Space Telescope Axial Replacement (COSTAR), effectively a system of mirrors which can compensate for the aberration caused by the faulty primary mirror. Astronaut Thomas Akers, lower left, is assisting Thornton to install COSTAR. The Hubble Space Telescope, with its instrument cluster doors open, can be seen in the background.

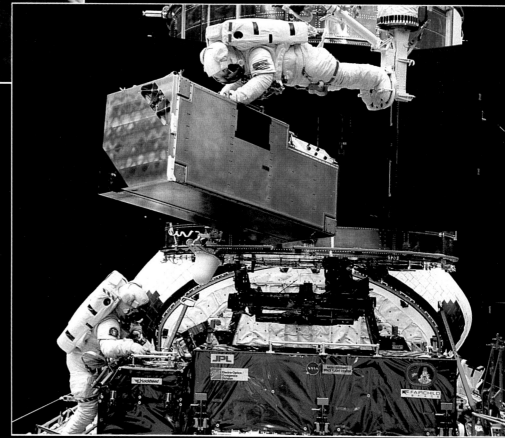

repair. The instrument to be sacrificed was the High Speed Photometer. The Wide Field and Planetary Camera would have to be entirely replaced with corrective optics of its own because it does not sit in the instrument cluster. The corrective unit itself was called the Corrective Optics Space Telescope Axial Replacement (COSTAR).

Detailed planning of the repair mission began in 1990, shortly after the problem with the mirror was discovered. As expected, several other problems arose and these were added to the list of tasks to be undertaken by the astronauts during what became known as Hubble's first servicing mission. The seven astronauts chosen for the mission were Jeffrey Hoffman, Thomas Akers, Dick Covey, Claude Nicollier, Story Musgrave, Kathryn

Endeavour's robotic arm *(left)* winches astronaut Story Musgrave to the top of the Space Telescope, where he will install protective covers over instruments designed to detect magnetic fields. This is one of a number of tasks that the astronauts on the first servicing mission had to perform in addition to repairing the mirror. Below Musgrave, in the cargo bay of the Space Shuttle, astronaut Jeffrey Hoffman can be seen assisting the repair efforts.

Thornton and Ken Bowersox. They practiced
for a year in order to be completely familiar
with the tasks they were expected to perform
in space. It is incredibly difficult to simulate
on Earth what the weightlessness of space
feels like, but a reasonably good idea can be
gained by working under water. This allows
movement through all three dimensions with
relative ease. The astronauts donned their
spacesuits and were submerged in a vast
swimming pool, known as a Neutral Buoyancy
Simulator. They trained with highly accurate
mock-ups of the equipment they would be
handling in space. Outside the pool, the
latest virtual reality software was developed
to simulate other aspects of the repair.

In the event, the first servicing mission was

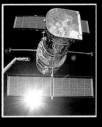

almost as if this type of mission was routine for NASA. It was a great achievement both for astronomy and for the public image of NASA itself. Following the completion of the mission, the really nerve-wracking part began; testing the efficiency of the repairs. Although many remained sceptical, by New Year's Eve 1993, there was no longer any shadow of a doubt: Hubble had been fixed. In fact, the results exceeded all expectations. Originally, Hubble had been designed to focus 70 percent of a star's light into a single point. Based upon the physics of optical systems, the best possible concentration of star light is about 87 percent. With the new optics in place, Hubble was surpassing its original specification (and even nudging the theoretical maximum) by focusing 84 percent of starlight into a central point. Almost overnight, it seemed, the telescope had become the finest instrument mankind had ever produced with which to study the universe. This view was celebrated on January 13, 1994 at a press conference when a plethora of images of galaxies and nebulae were presented to the eager members of the press. The Hubble Space Telescope, it was proclaimed, would allow astronomers the opportunity to practise 21st century science several years before the turn of the millennium.

A steady improvement in resolution *(right).* Reading from the top, the first image shows a region within galaxy M100, taken by the 200-inch (5-meter) ground-based telescope on Mount Palomar. Below it, is the same region taken with Hubble's first Wide Field and Planetary Camera. Even with aberrated optics, it is still a better image than can be taken from the ground. The image below that is a computer processed version of the same shot. Following repair, a much clearer picture can be achieved with the new Wide Field and Planetary Camera (bottom image). The arrows point to Cepheid variable stars, used by astronomers to calibrate distances in the universe.

Images from the Wide Field and Planetary Camera *(right and far right)* before and after the Hubble repair mission. The image (right) was taken with the original Wide Field and Planetary Camera just days before it was replaced. Although it was an excellent image by ground-based standards, it is completely surpassed by the image (far right) which shows the same view taken by the second Wide Field and Planetary Camera with corrective optics incorporated in its imaging system. In the second view the central region of the spiral galaxy M100 is extremely well defined. This was one of the images that finally convinced Hubble scientists that the corrections to the optical systems had been successful. It was released to the public on January 13, 1994 at a celebratory press conference.

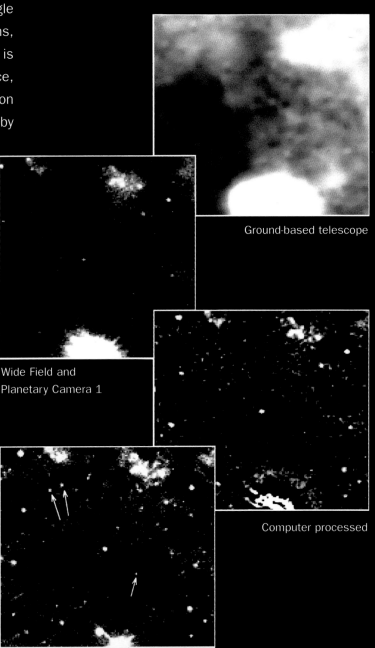

Ground-based telescope

Wide Field and Planetary Camera 1

Computer processed

Wide Field and Planetary Camera 2

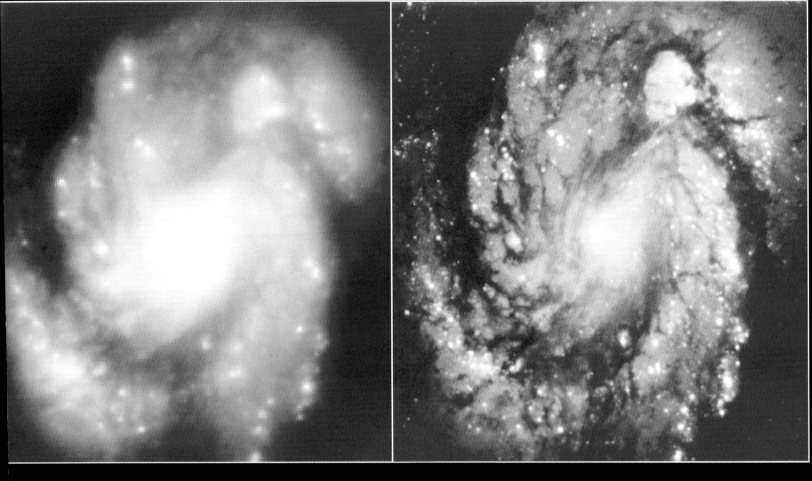

Before and after COSTAR *(below)*. The two spectrographs and the Faint Object Camera at the base of the Hubble Space Telescope were not replaced during the repair mission. Instead, the High Speed Photometer was replaced by the COSTAR module, which was added to refocus the light for the three remaining instruments. The image (left) taken by the Faint Object Camera before repair is abberated; the image (right) taken after repair, when the Faint Object Camera was used in conjunction with the COSTAR module, shows a significant improvement. Detailed analysis has shown that Hubble is now working at the very limits of what theory says is possible for a 95-inch (2.4-meter) telescope.

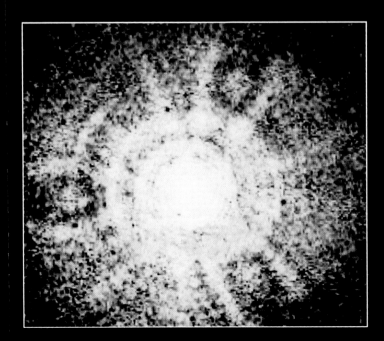

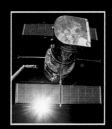

HUBBLE'S IMAGES

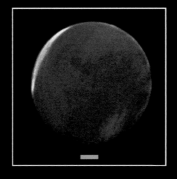 

THE HUBBLE SPACE TELESCOPE is so powerful that, if it were above Washington D.C., it could see a firefly in Tokyo! If there were two fireflies in Tokyo, Hubble would be able to tell them apart, so long as they were separated by just ten feet (three meters). Helping Hubble to achieve the equivalents of these feats in space, are the onboard cameras and spectrographs. At present, these instruments work around the visible region of the spectrum. Short visible wavelengths are seen as blue light and long visible wavelengths are seen as red light. Radiation with wavelengths shorter than blue light are known as ultraviolet rays and those longer than red light are known as infrared. Hubble's instruments can detect all of the visible spectrum, some ultraviolet and a little of the infrared (known as "near-infrared" because it is are so close to red light). Observing the universe at different wavelengths provides astronomers with a view of a number of different physical processes. In time, each generation of instruments will be replaced by more sophisticated ones, which will allow the detection of more ultraviolet and, especially, infrared radiation. The second Hubble servicing mission, successfully completed in February 1997, involved installing two new instruments: the Near-Infrared Camera and Multi-Object Spectrometer (NICMOS); and the Space Telescope Imaging Spectrograph (STIS), designed to span ultraviolet, visible and infrared wavelengths.

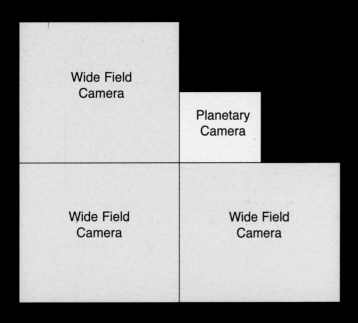

How color images are made *(above)*. CCD cameras are not particularly sensitive to the color of light, and can only produce a color picture by combining three or more separate images. To produce a true-color image, one picture is taken through each of three different filters (red, green and blue). This sequence of images shows the planet Mars through red, blue and green filters, and a resulting true-color image which is a computer-processed composite of the three.

Hubble's Wide Field and Planetary Camera *(left)* is a composite of four CCDs; three wide-field CCDs and one planetary CCD operating at higher magnification. More detail can be seen in the planetary frame but, when data from all four CCDs are displayed together, the planetary image must be reduced so that it is at the same scale as the others. This produces a "stepped" image as seen on pages 95–97.

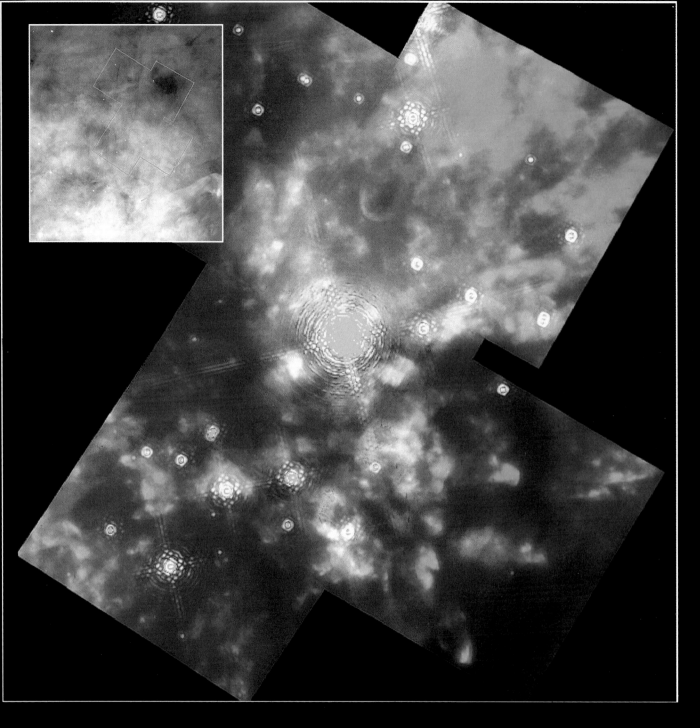

All Hubble's instruments (both current and proposed) use electronic cameras known as Charge Coupled Devices (CCDs) to record the information for transmission back to Earth. CCDs are light-sensitive computer chips that contain a two-dimensional network of microscopic "buckets". These buckets are known as picture elements (pixels for short) and convert any light that strikes them into an electric current. The strength of the

A visible and infrared view of the Orion nebula *(above)*.
Within the outlined area of the visible light image (inset), taken with the Wide Field and Planetary Camera (WFPC), little is visible because the glowing gas cloud also contains dust, which blocks our view of the stars. However, in the infrared image of the same area, taken by the Near-Infrared Camera and Multi-Object Spectrometer (NICMOS), young stars and warm dust can be seen as orange/yellow objects, since infrared radiation can pass more easily, than visible light, through the cloud. The blue streamers and arcs are high speed jets of gas, colliding with slower-moving gas.

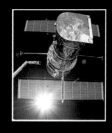

electric current in each pixel is quite easily transformed into computer data, which determines how bright the pixels appear when displayed on a computer screen on Earth. If the Hubble Space Telescope is looking at a nearby object or, indeed, at a very large one farther away, it produces a high-resolution image in which the individual pixels are almost impossible to see. If, however, the telescope is looking at a very small object or something incredibly far away, the technicians processing the image often have to make enlargements. Once this is done, the pixels may become visible and the image looks as if it is composed of a grid of colored boxes. The result is known as a pixelated image (the first impression of Pluto on page 31 is a good example).

Storing the images as computer data means that they can be manipulated using image-processing techniques. Technicians can accentuate contrasts in the images, so that faint details can be brought to the fore. Another technique used to highlight detail is the creation of a false-color image. Colors are imposed during processing to provide striking contrasts between one aspect of the image and another. False color may also be imposed to pick out different chemical compositions or different temperatures of gas. (A typical example can be seen on page 41, Saturn's rings).

Hubble's view compared to the view from a ground-based telescope *(right)*. Hubble's major advantage is that it is above the Earth's atmosphere, which distorts the images received by ground-based telescopes, even when optimum weather conditions prevail. Unhindered by this problem, Hubble sees many objects that are beyond the range of even the largest ground-based telescopes. It does this in spite of the fact that its mirror is not as large as some of the ground-based telescopes.

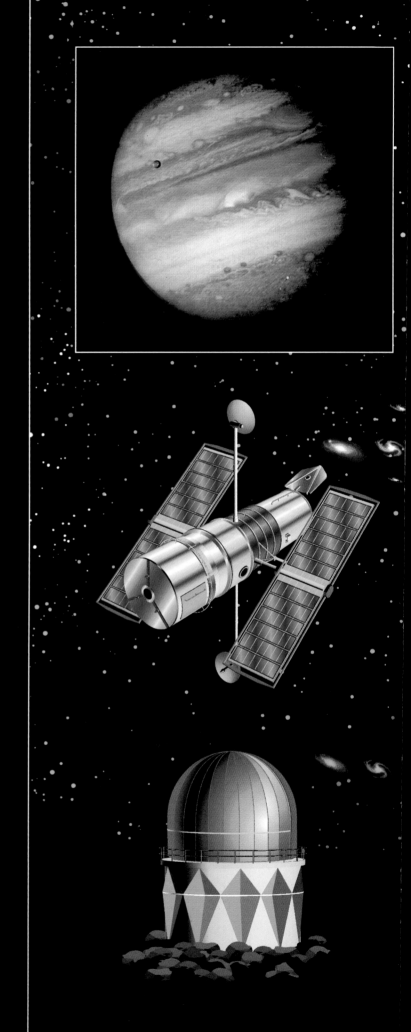

Jupiter photographed by Voyager 1
(left) during the late 1970s. The probe
passed through the Jovian system with-
out stopping, but provided scientists
with the clearest images of Jupiter ever
seen at the time. The disadvantage
was that it could not observe
the planet in the long term.

Jupiter photographed by Hubble and from the ground
(above and right). In the past, ground-based telescopes were
used to monitor Jupiter and returned many photographs like
the one shown (right). More recently, Hubble has proved that
it can take sharp planetary images (above), of a quality
comparable with those obtained by the Voyager space probes,
but it can do so repeatedly rather than being restricted to
taking a small number of photographs on a fleeting visit.

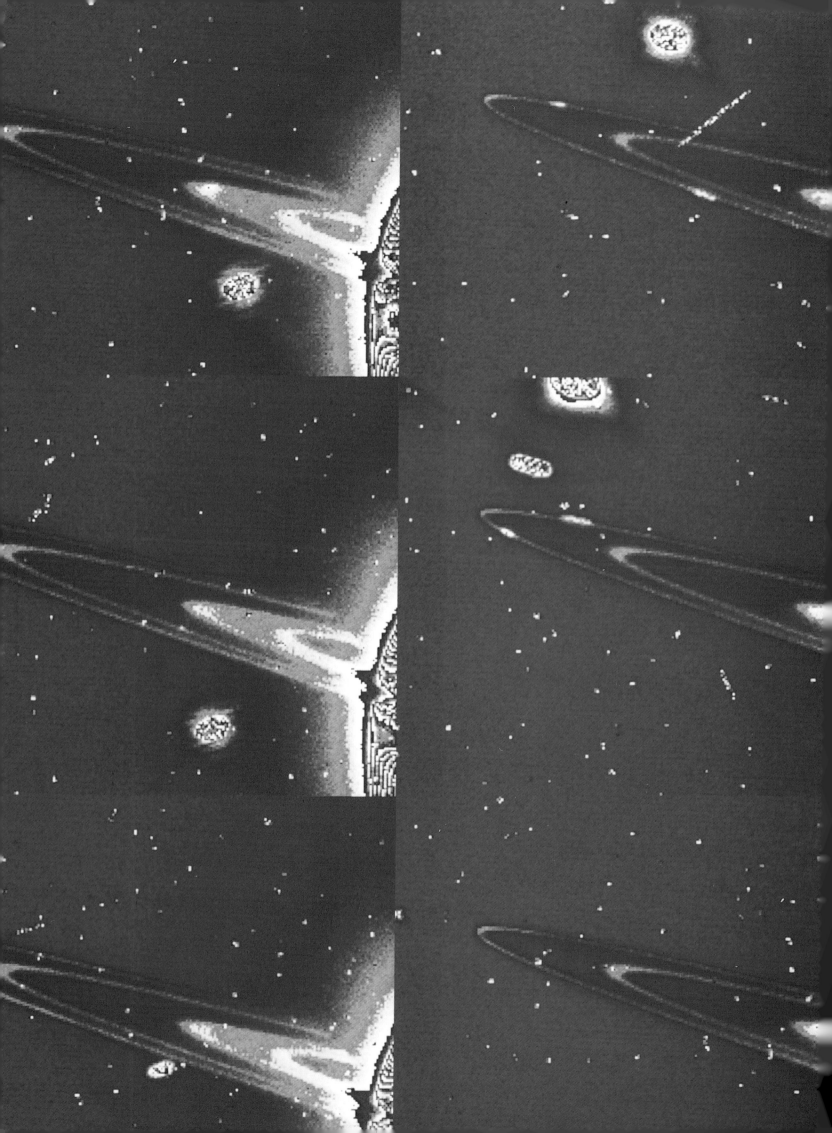

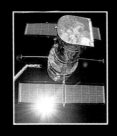

THE
SOLAR
SYSTEM

THE EARTH IS ONE OF NINE PLANETS THAT ORBIT THE SUN. These planets form the major part of the Sun's family, or Solar System, which also includes a myriad of moons and other smaller bodies, such as comets and asteroids. The Hubble Space Telescope has begun to study these, our celestial neighbors, in unprecedented detail, providing an abundance of valuable new information. It has also spearheaded the search for new objects in the outer reaches of the Solar System.

DISCOVERY OF THE PLANETS

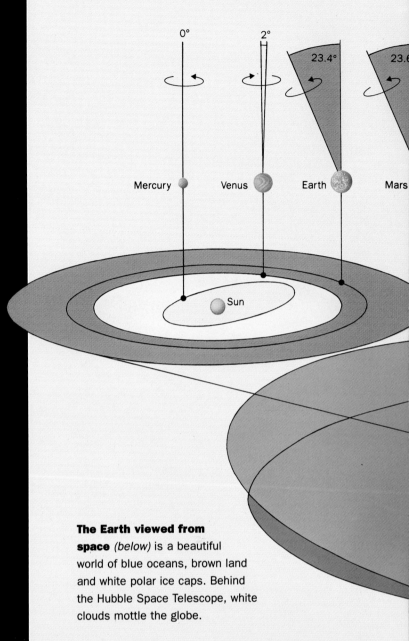

Mercury   Venus   Earth   Mars

0°   2°   23.4°   23.6

Sun

SINCE THE EARLIEST CIVILIZATIONS, mankind has aspired to an understanding of the stars. The ancients imagined that the brightest stars were great heroes and villains, and told stories about the adventures of these mythological creations. They knew of five "wandering stars," brilliant lights that never twinkled like the others, and always followed the same path across the sky. The Sun and the Moon also inscribed the same route across the sky.

Today these "wandering stars" are known to us as the other planets in our Solar System (the word "planet" is from the Greek for wandering star). Five of them, Mercury, Venus, Mars, Jupiter and Saturn, can be seen easily with the naked eye. Using a telescope, in 1781 the British astronomer William Herschel discovered the planet Uranus. In the years that followed, continual study of Uranus showed that it was being pulled by the gravity of another, more distant world. Making use of these calculations to narrow the search area, a new quest was undertaken. Neptune was finally identified in 1846 by the German astronomer Johann Galle. In 1930 yet another planet was found when the American, Clyde Tombaugh, discovered Pluto. Some astronomers believe that there is a tenth planet, out beyond Pluto, which is still awaiting discovery.

The Earth viewed from space *(below)* is a beautiful world of blue oceans, brown land and white polar ice caps. Behind the Hubble Space Telescope, white clouds mottle the globe.

Schematic of the Solar System *(above)* showing the orbits of the planets and giving some indication of their relative sizes (only a portion of the Sun, represented by the orange area, can be shown at this scale). The planets are all confined to the same orbital plane, known as the ecliptic, except for Pluto, which is slightly inclined. Each of the nine planets rotates about its axis and, for most of them, the direction of rotation is west to east. The exception is Venus, which rotates east to west. Apart from Uranus, which appears to be on its side, the planets' rotation axes are, more or less, at right angles to the ecliptic. Most of the orbits are nearly circular in shape but Mercury, Mars and Pluto are noticeably elliptical. The extent of a planet's deviation from a circular orbit is known as its eccentricity. The planets of the Solar System can be grouped into the four inner rocky planets and the four outer gas giants. This leaves Pluto as the "odd planet out".

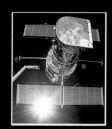

ROCKY
AND ICY
WORLDS

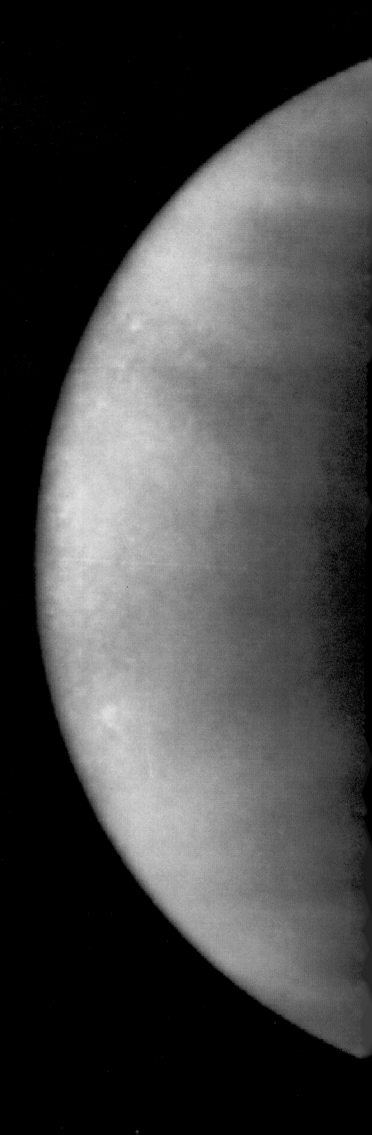

FOUR ROCKY PLANETS, including Earth, comprise the inner Solar System. The innermost planet, Mercury, is a barren place that bears a superficial resemblance to Earth's Moon. Because of its proximity to the Sun, it is constantly bathed in fierce heat and radiation. As yet, Mercury has not been imaged by the Hubble Space Telescope. The next nearest planet to the Sun is Venus. In some respects Venus can be thought of as a twin to our home, Earth. Specifically its mass is very similar to that of Earth but, because it is closer to the Sun, Venus has evolved along very different lines. Space probes have shown that Venus was once very active geologically, and experts believe that volcanoes once pumped vast quantities of gas into the planet's atmosphere. Because Venus lacked plant life to regulate the build-up of these gases, they now insulate the planet and act like a greenhouse. Radiation from the Sun streams onto the planet by day and heats the surface. At night that

Venus shows some interesting cloud features *(main image and top right)* when viewed in ultraviolet light. The clouds are composed of sulfuric acid rather than water vapor and the dark, "V"-shaped features near the equator show concentrations of sulfur dioxide. These features are seen most easily in the image taken at relatively close quarters by the space probe Pioneer Venus (top right), but can just be seen in the shot taken by Hubble (main image). Toward the planet's poles, the clouds begin to follow the lines of latitude. Like our own Moon, Venus appears to go through phases and the only way to capture fully global pictures is to send space probes to the planet.

heat is re-radiated but in a form of energy that cannot escape the dense, all encompassing, cloud layers. By this process, the surface of the planet has reached a scorching 480°C, making Venus the hottest planet in the Solar System.

This is in stark contrast to Pluto, the coldest and smallest planet in the Solar System, found on the very fringes of the Sun's domain. Pluto is so far away from the Sun that the chemical methane, which covers it, has been frozen to its surface forming a highly reflective layer of ice. The surface of the planet is a frigid −200°C. There is some contention among astronomers over whether Pluto should have been classified as a planet at all. One opposing view is that Pluto may simply be the largest of a group of icy asteroids that exist in the farthest reaches of our Solar System. Several other icy objects are known to exist in this region, and it is part of the Hubble Space Telescope's mission to search for more.

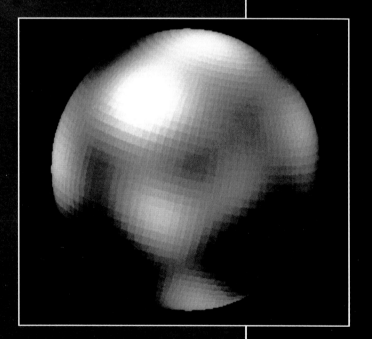

First impressions of Pluto's surface *(above)* were revealed in this Hubble image. Each pixel represents an area on Pluto's surface over 100 miles (160 km) across. Using other shots like this, astronomers have been able to combine the images to produce a map of the surface *(above left)*. This is an astounding feat, considering that Pluto was 3 billion miles (4.8 billion km) away from Hubble when the image was taken.

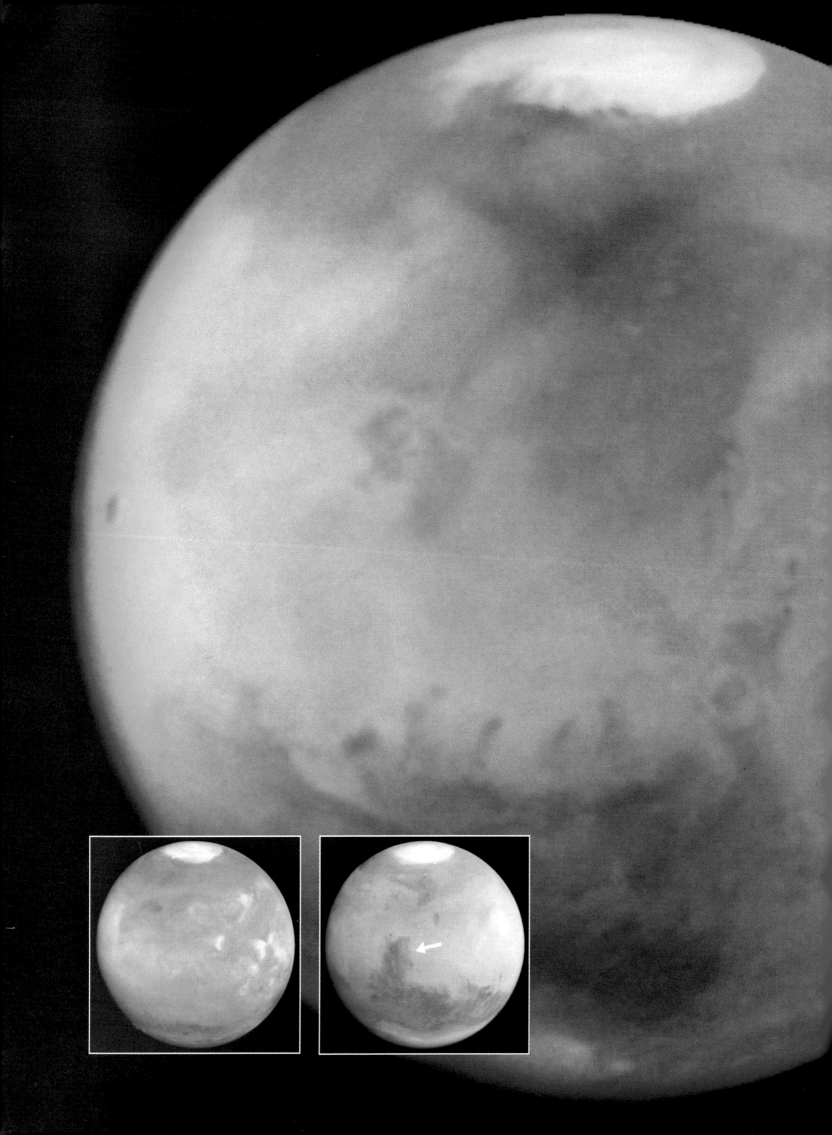

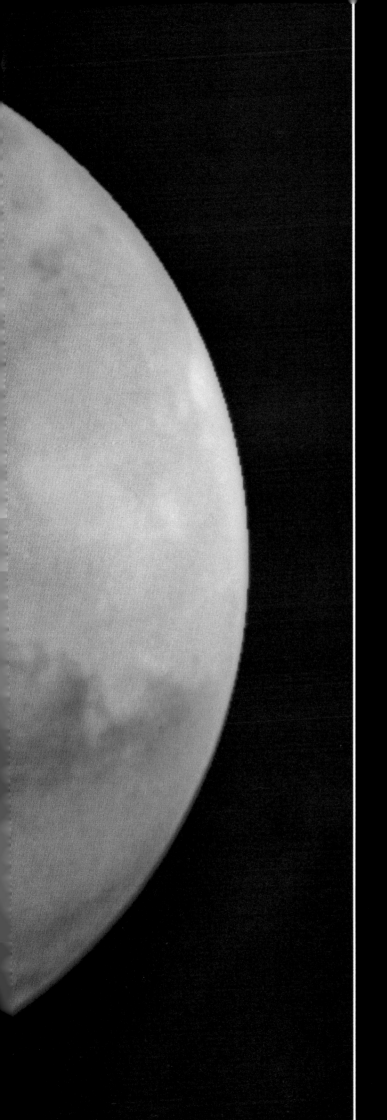

Mars is the fourth planet in the Solar System and the only one on which surface features can be viewed by ground-based telescopes. The first observation of Mars took place in 1659 by Dutch physicist Christian Huygens using a home-made telescope. He identified a dark triangular feature, known today as Syrtis Major, which he observed for several weeks, concluding that a Martian day was similar to our own – about 24 hours long. The spectacular red of the Martian surface led some early astronomers to believe that the planet was covered in an alien vegetation. However, space probes have proved beyond doubt that the color is, in fact, produced by the red dust that covers the planet. They have also found evidence that running water once flowed on Mars. This was a particularly exciting discovery, because it raised the question why did the Martian atmosphere gradually lose its water but the Earth's did not? One possible answer is that Mars had no ozone layer. On Earth this protects water molecules (and life forms) from harmful ultraviolet radiation from the Sun. On Mars the water molecules were broken up by that radiation; the constituent hydrogen (being very light) drifted into space and the oxygen sank to the surface. On the surface, oxygen combined with metals in the rocks, oxidizing them and turning them rusty red. Mars is a geologist's paradise, just waiting to be explored. In particular, it is the site of the largest known volcano in the Solar System, the extinct Olympus Mons. It will almost certainly be the next planet to which we will send astronauts.

The rift valley, Valles Marineris, on Mars *(left)* appears as the dark blue region across the lower half of the image. The brown spot, poking through the cloud on the left, is the extinct volcano, Ascraeus Mons. Its effect on the atmosphere, and that of its four "brothers", can be seen in the Tharsis region *(inset left)*. The white crescents are caused by warm air pushed upward by the volcanoes to form ice crystals down-wind of the summits. The low-lying plain, Syrtis Major *(inset right)*, can be seen as a dark protrusion near the center of the image, marked with an arrow.

THE GAS GIANTS

PROGRESSING OUTWARD in the Solar System, the next destination is mighty Jupiter, largest of the nine planets. Although the Sun accounts for over 99 percent of the mass in the Solar System, Jupiter embodies about 70 percent of what remains. Its diameter is over 11 times that of the Earth. Jupiter is mostly composed of gas, particularly hydrogen and helium. It is made of exactly the same material as the Sun, in exactly the same proportions; had it been much larger, nuclear fusion would have started in its core and it would have become a star.

Jupiter has no solid surface. Its gaseous atmosphere simply becomes denser and denser, transforming first into a liquid and then into a metallic liquid, similar to the chemical mercury. Buried deep in the center of this enormous planet may be a rocky core some 13 times the mass of the Earth, but this theory has not been tested. Jupiter is banded with clouds, and what we see when Hubble looks at Jupiter is the topmost layers of cloud. The clouds follow lines of latitude and are called by different names, depending upon whether they are dark or light in color. Dark cloud bands are known as belts and the light bands are called zones. The zones are thought to be high level clouds that catch the Sun's light and reflect it to Earth; belts are imagined to be lower cloud decks that appear to be in shadow. There is a large oval in Jupiter's atmosphere, which is easily visible with a telescope from Earth and has been named the Great Red Spot. It is actually a tornado-like storm, which has been observed since the mid-1600s. It

spins anti-clockwise and takes just under one week to complete one revolution. Time-lapse photography has shown that the storm is kept active by powerful opposing air currents set up because the winds to the north of the spot blow due west, while those to the south blow east. Smaller, and much shorter-lived spots have also been observed in the planet's atmosphere at the boundary of two air currents moving in opposite directions.

Views of Jupiter *(right and below)* taken by Hubble as part of the preparation for a mission by the NASA space probe, Galileo, currently monitoring Jupiter. As part of its mission, Galileo deployed a smaller probe to plunge into Jupiter's atmosphere and sample its composition before being crushed by the titanic pressure. Hubble took this image of Jupiter (right) on October 5, 1995, two months before the probe entered Jupiter's atmosphere on December 7, 1995 (the black arrow marks the probe's actual entry point). The sequence of four enlargements (below) show atmospheric change around the probe's destination (a white dot has been superimposed on the entry point). Each image spans an area of Jupiter three times the diameter of the Earth. The image on the left was taken on October 4, 1995 and (reading left to right) the others were obtained 10, 20 and 60 hours later. In this region of the planet's atmosphere, winds blow at speeds of around 250 miles per hour (110 meters per second). In the time elapsed between the first and fourth image, the winds have swept the cloud formations some 15,000 miles (24,000 km) westward. It is intended that Hubble should continue to support other NASA missions in this way.

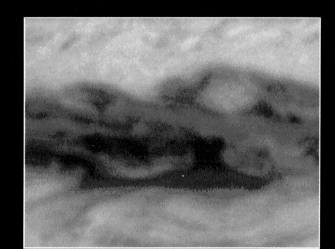

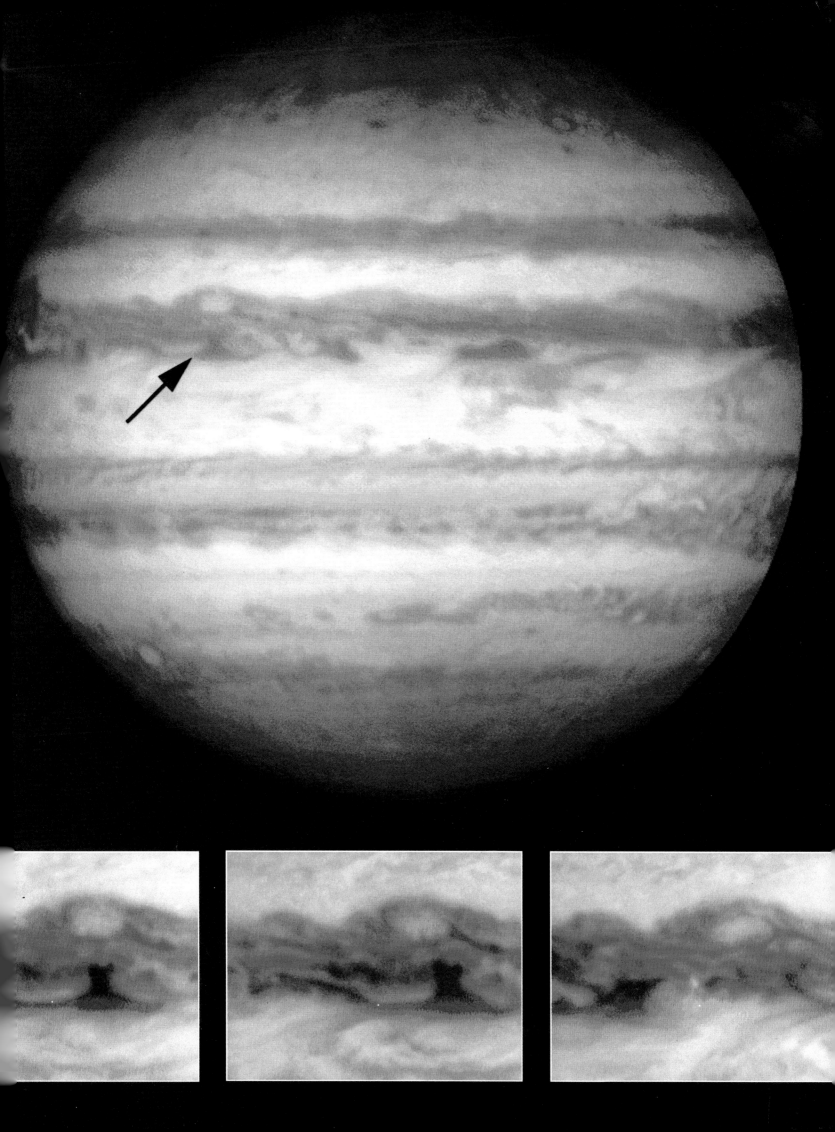

This fascinating world was first studied with a telescope by Galileo Galilei in 1610. He discovered four large moons which he named the Medicean Stars. They have since been given individual names, Io, Europa, Ganymede and Callisto, and are jointly referred to as the Galilean satellites of Jupiter. Despite the fact that they are all moons of the same planet, these four worlds are very different from each other. Io is the most volcanically active body in the Solar System. It is in a constant state of eruption, with volcanoes spewing molten sulfur onto its surface all the time. The volcanic activity on Io is caused by the titanic pull of Jupiter's gravity which squeezes the little moon and causes its interior to melt through friction. Europa, although not as volcanically active as Io, is also heated by the pull of Jupiter's gravity. In this case the heating may have melted ice producing underground reservoirs of liquid water, which is a particularly exciting concept because they would provide possible homes for primitive, microbial life. Ganymede and Callisto are cold, icy and barren places. They show quite heavily cratered surfaces that appear to have remained largely unchanged since the final stages of the Solar System's formation, roughly 4.5 billion years ago. Although all four of the Galilean satellites are big, Ganymede and Callisto are particularly large in comparison with most other moons in the Solar System. They are icy bodies with cratered surfaces and are almost the same size as the planet Mercury. As well as these large moons Jupiter also possesses at least 12 other, much smaller moons. Four of them are inside the orbit of Io and the remaining eight are found beyond Callisto. They are all tiny rocky bodies, similar to asteroids.

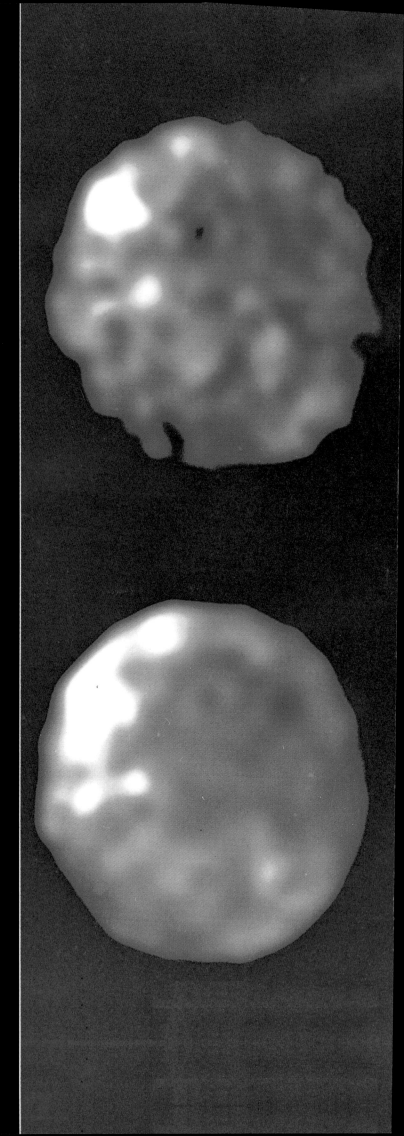

Jupiter's moon Io *(left)* photographed in four different ways to allow astronomers to examine different facets of one of the Solar System's most intriguing objects. The visible light photograph (top left) was taken in March 1992 before Hubble's optics were corrected. The image (top right) was also taken by Hubble, but in ultraviolet light. Although the two pictures show the same face of Io, they look very different. Ultraviolet light picks up regions of different temperature and composition. The visible light image (bottom right) was taken by the space probe Voyager in 1979. To look for changes on Io during the intervening years, the Voyager image was processed (bottom left) to match the resolution of the Hubble shot. Comparison between the two shows that the surface of Io has changed considerably over the past several years.

The moons of Jupiter *(below)* can be imaged in considerable detail using the Hubble Space Telescope's corrected optics. During 1995, Hubble has charted new volcanic activity on Io (top left), identified ozone on Ganymede (top right), discovered a thin oxygen atmosphere on Europa (bottom left), and viewed fresh ice deposits on Callisto that indicate meteorite impacts (bottom right).

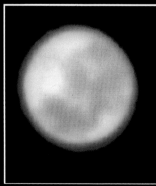
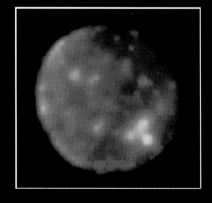

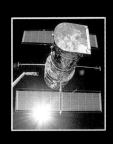

Saturn and its rings *(right)* in a false-color image that is unusual because of the way the rings are illuminated by the Sun. Usually, we see sunlight reflecting off the densest rings (A and B) because the Earth and the Sun are slightly above the plane of the rings. In this shot, the Sun is below the ring plane and so sunlight most easily reaches the Earth through the rings that contain the least amount of material. In the image, the brightest ring is the outermost F-ring, next brightest is the Cassini division and the third brightest is the C-ring.

The sixth planet of the Solar System is the spectacular ringed world of Saturn. This magnificent planet is almost twice as far away from the Sun as Jupiter, and slightly smaller, but shares many of its characteristics. The clouds in Saturn's atmosphere display a banded structure similar to that of Jupiter, but the patterns and colors are not quite so striking. This is probably because Saturn does not generate as much internal heat as Jupiter. Very detailed imaging of the planet shows vortices and storm fronts at the boundaries of zones and belts. Periodically, the yellow hues of Saturn's atmosphere are heavily disrupted by the outbreak of great storms that sweep across the equatorial regions of the planet pushing highly reflective ammonia crystals into the upper atmosphere. The last such storm occurred on the planet in September 1994, and was captured by Hubble. Astronomers believe that, below the visible cloud tops, a curious phenomenon is

The great storm in Saturn's atmosphere *(above)* during December 1994. The area of intense activity appears as a white shape, like a shark's dorsal fin.

taking place. The conditions in the planet are considered to be just right for helium droplets to condense and fall like rain to the lower levels of the planet. This process is thought to have begun nearly two billion years ago and is still continuing today. In other words, it has been raining helium on Saturn for almost half the age of the Solar System. The most distinctive feature of Saturn, however, is its glorious ring system. It is visible from Earth, even with a small telescope, and was initially thought to consist of two separate rings, named A and B,

and separated by a division named after its discoverer, the astronomer G. D. Cassini. By the mid-1800s a fainter inner ring had been discovered, now called the C-ring. As space telescope technology has improved so the structure of Saturn's ring system has been revealed to be ever more complex. There are now seven known components to the ring system. The rings

themselves are composed of tiny dust particles and pebbles with, perhaps, a few rocks and small boulders. The innermost ring is the D-ring. It is very faint and was discovered by the Voyager space probes as they passed the planet during the early 1980s. It extends from the cloud tops to the edge of the C-ring. The C-ring itself extends out to the edge of the B-ring, the brightest of Saturn's rings when viewed from Earth. Beyond the B-ring lies the Cassini division and then the A-ring. Beyond this are three extremely faint rings known as F, G and E, named in order of discovery. These three outer rings are held in place by gravity from the tiny moons of Saturn, known as shepherd satellites.

Like Jupiter, Saturn has a cluster of moons. Of those discovered so far, 18 are considered to be prominent; one of them, Titan, is the second largest moon in the Solar System (after Jupiter's Ganymede). Titan is swathed in a dense atmosphere, which some scientists believe resembles the composition of the Earth's atmosphere four billion years ago, before life began. Titan is so far away from the Sun that it lacks the energy thought to be necessary for life to emerge. Saturn's other satellites are small worlds of rock and ice and each one has a distinctive character all of its own.

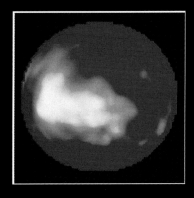

Ultraviolet image of Saturn *(right)* showing aurorae, which are produced where tiny particles from the Sun have slammed into Saturn's upper atmosphere.

Titan, Saturn's largest moon *(above)*. In this unprecedented image Hubble peered through the haze surrounding the moon and glimpsed a large bright structure, never seen before. It is about the size of Australia, but astronomers do not know whether it is land or ocean.

Saturn's rings seen edge on *(below and bottom)* during the "ring plane crossing" in May 1995, when the plane of Saturn's rings was level with our line of vision. Hubble discovered a new moon, S/1995 S3, visible as the smudge at the center of the picture (below). Crossings are rare but useful because, with the glare from the rings temporarily dimmed, astronomers can study Saturn's moons more easily. Several moons are visible in one shot (bottom) where the ring is almost level with the camera. The shadow on Saturn's disk is cast by Titan.

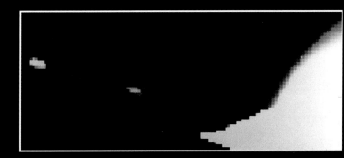

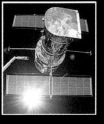

The other gas giants in the Solar System are Uranus and Neptune. Although both are smaller than Jupiter and Saturn, their diameters still measure about four times that of the Earth. They inhabit the distant reaches of the Solar System, far away from the Sun, where heat and light are very weak. In these conditions, gases such as methane and ammonia begin to condense into ice. On Uranus and Neptune, the blue color of each world is produced by the light-absorbing properties of this ice (especially methane) in the planets' atmospheres. Uranus is a fascinating world because its rotation axis is so highly inclined towards the ecliptic. The north pole of the planet rises above the plane of the Solar System by 7.9 degrees. Uranus is the only planet that receives more sunlight at its poles than at its equator, because, at only two points during its orbit of the Sun, is the equator closer to the Sun than one or the other pole. Uranus has no internal heat source and is, to most intents and purposes, inert. As a result, the atmosphere is virtually featureless, although careful image processing can show a hint of banded cloud structure stretching around the planet. Neptune, despite being twice as far as Uranus from the Sun's heat, possesses a dynamic atmosphere with cloud belts and cyclonic storms. This is almost certainly because Neptune, like Jupiter and Saturn, has an internal heat source, which creates atmospheric change producing storms and different weather conditions. When the Voyager 2 space probe passed Neptune in August 1989, it captured pictures of a spectacular cyclone, which became known as the Great Dark Spot. It was a dark blue oval, so large that the entire Earth could have fitted into this tornado-like feature of Neptune's atmosphere. During the years that it has been in orbit, the Hubble Space Telescope has kept a careful watch on Neptune, charting the changes in its turbulent atmosphere.

Both Uranus and Neptune have interesting families of satellites. Uranus has at least 15 moons, 10 of which are very small and were found by the Voyager 2 space probe during its fleeting visit in 1986. Of the five large moons, most are icy environments; for example, Miranda possesses some gigantic ice cliffs, some of which are 12.5 miles (20 km) high and Ariel displays some spectacular ice-rift valleys. Neptune has at least eight satellites. Of these, Triton is the most fascinating because its surface is pock-marked by strange geysers that spit liquid nitrogen into the tenuous atmosphere of the moon. The nitrogen migrates downwind for quite a distance, before finally settling back on the surface.

The changing face of Neptune *(right)* due to its turbulent atmosphere. From left to right, the images were taken on October 10, October 18 and November 2, 1994. The clouds are composed of high-altitude methane crystals and are tinged pink because Hubble imaged the planet at visible and near-infrared wavelengths. The Hubble Space Telescope now keeps a regular check on the atmospheres of the outermost planets so that astronomers can keep track of their constantly changing weather patterns.

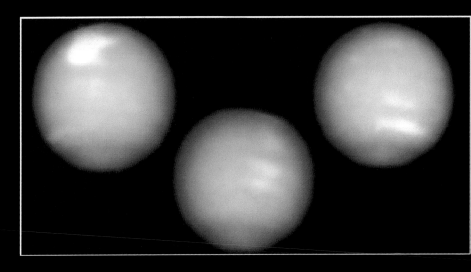

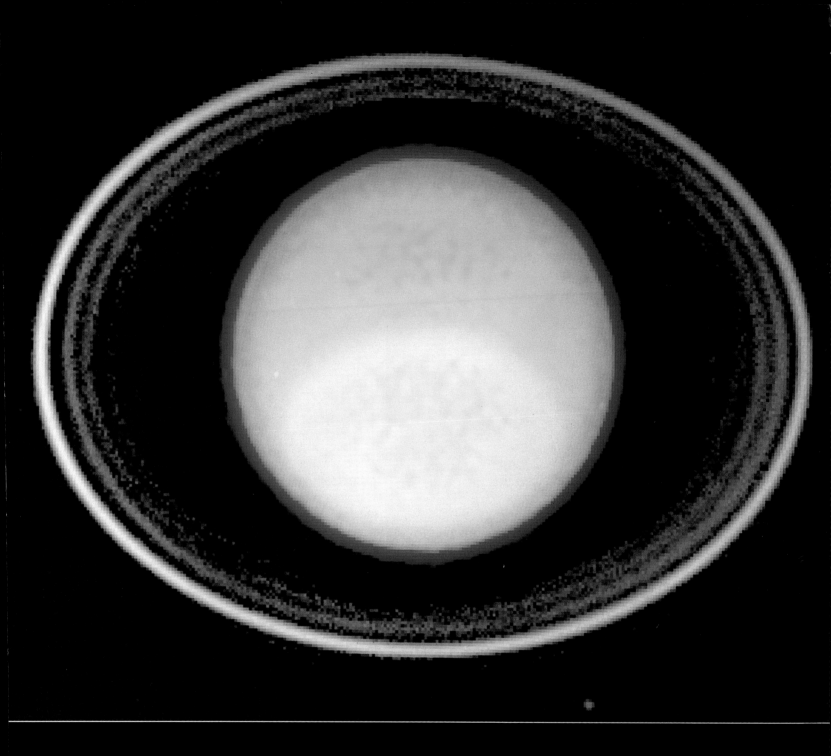

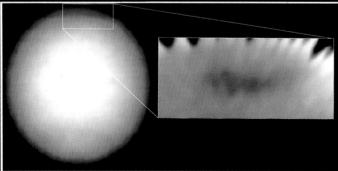

A second Great Dark Spot on Neptune *(above)*, similar to the one that Voyager found, but in a different hemisphere of the planet. Hubble will track it carefully, watching for changes.

Uranus pictured in April 1996 *(above)* using detectors working in the near infrared, a technique that was not available to the Voyager 2 space probe of 1986. This image has allowed astronomers to perform detailed chemical analysis of the Uranian atmosphere. The picture has been falsely colored to show high-altitude haze in pink and dark blue. Lower-level methane-rich clouds are shown in light blue and yellow. Image processing has also been used to brighten the rings around the planet because, in reality, they are composed of material that is as black as charcoal. The rings loop around Uranus above its equator, allowing astronomers to work out that we are looking at the planet's south pole.

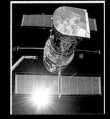

COMETS AND METEORITES

THE SOLAR SYSTEM does not just comprise nine planets, it is also made up of a host of tiny rocky and icy objects known as the minor bodies of the Solar System. Among them are rocky asteroids, icy comets and meteorites that fall to Earth. The minor bodies have remained virtually unchanged since the formation of the Solar System, 4.5 billion years ago and are much sought after for study. Comets are usually seen from Earth as ghostly white objects with expansive tails stretching through space. The tails are produced when the comet's orbit takes it into the inner Solar System where heat from the Sun melts the ice on the comet's surface. The melted ice becomes a gas that billows into space, forcing dust and other solid particles to be ejected, and leaving a trail of debris behind it. The nucleus may resemble a dirty iceberg about 12.5 miles (20 km) in diameter, composed of a mixture of small rocks and frozen gases.

Hubble discovers a tiny cometary nucleus *(above, tinted yellow)* in the outer reaches of the Solar System. It is one of only 29 such objects identified so far. Experts believe that there are about 200 million of them beyond Pluto.

Comet Hale Bopp *(right)* is a bright new long-period comet. Hubble captured this image on September 26, 1995 during the comet's plunge toward the inner Solar System. The stars around it appear to be tiny streaks of light. This is because the comet is moving so fast that Hubble needed to track it in order to produce a sharp image.

Comets that venture into the inner Solar System can be divided into two types, long-period and short-period. If a comet completes its orbit in less than 200 years, it is termed a short-period comet. A prime example is Halley's comet, which returns to the inner Solar System every 78 years. Long-period comets typically take between 100,000 and 1,000,000 years to complete one orbit, and travel some 45,000 times the distance of the Earth's orbit into parts of the Solar System away from the Sun. Most comets are believed to be beyond Pluto, but finding examples so far out in space is

Shoemaker-Levy 9 *(right and insert).* The main image shows fragments of the comet in close-up while the insert uses superimposed images to create a snapshot of the comet fragments as they were about to collide with Jupiter. First Hubble took a picture of Jupiter using a short exposure (careful inspection of Jupiter will reveal the moon Io and its shadow, passing across the disk of the planet). Then Hubble captured the much fainter fragments using a longer exposure, and technicians joined the frames together. At the time this image was taken, the 21 fragments of Shoemaker-Levy 9 stretched across 700,000 miles (1.1 million km) of space and were mere days away from collision.

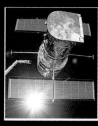

very difficult. Only with instruments like the Hubble Space Telescope can astronomers hope to glimpse these tiny objects. Being so small, comets are easily affected by the gravitational pull of the planets. Jupiter distorts some comet orbits to the extent that they never return to the inner Solar System.

In March 1993, a very peculiar comet was discovered which had been captured by the gravitational field of Jupiter. Known as Shoemaker-Levy 9, it had been pulled to pieces by the gravity of Jupiter and the fragments were on a collision course with the planet. Although the Galileo probe was en route to Jupiter, it would not arrive in time to photograph the collision 15 months later. The Hubble Space Telescope became the principal instrument observing the spectacular event. As each of the icy fragments plunged into Jupiter's atmosphere the disturbances in the cloud base were captured by Hubble's cameras. The collision caused plumes of dusty material to be pushed into the atmosphere. These appeared on the Hubble images as

Jupiter's atmosphere pockmarked by impact sites *(right)*. They appear as dark brown areas on the far side of the main image. The diagram shows Hubble's position in relation to the impact sites when the images (inset) were taken. (1) At the point of impact, Hubble saw the top of the mushroom cloud explosion as it rose over the limb of the planet. (2) A few minutes later the debris had begun to flatten and spread out. These first two images were taken in the near infra red, capturing hot gas at the top of the mushroom cloud as it broke through the shadow on the far side of the planet. (3) Ninety minutes later, the impact site had rotated into daylight and, through clouds, Hubble took shots of the aftermath of the collision. The debris formed a horseshoe shape surrounded by a shockwave. An earlier impact site is to the left of the shockwave. (4) Days later atmospheric movement had broken up the pattern.

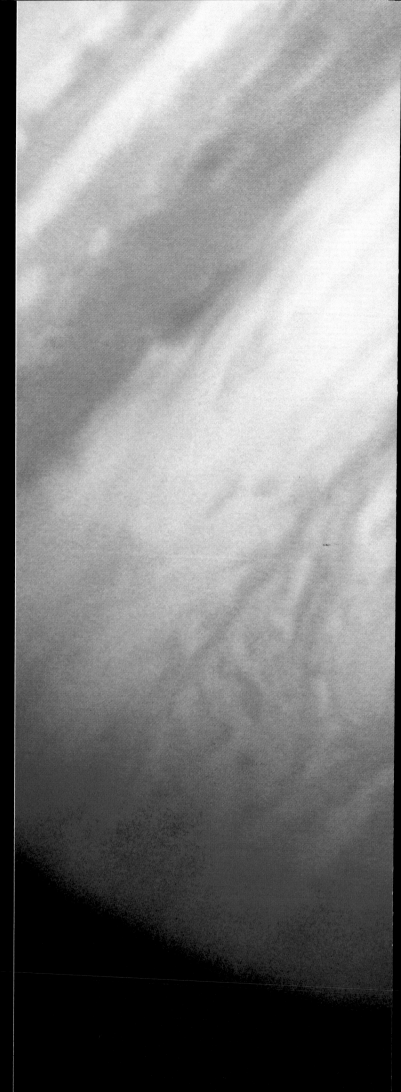

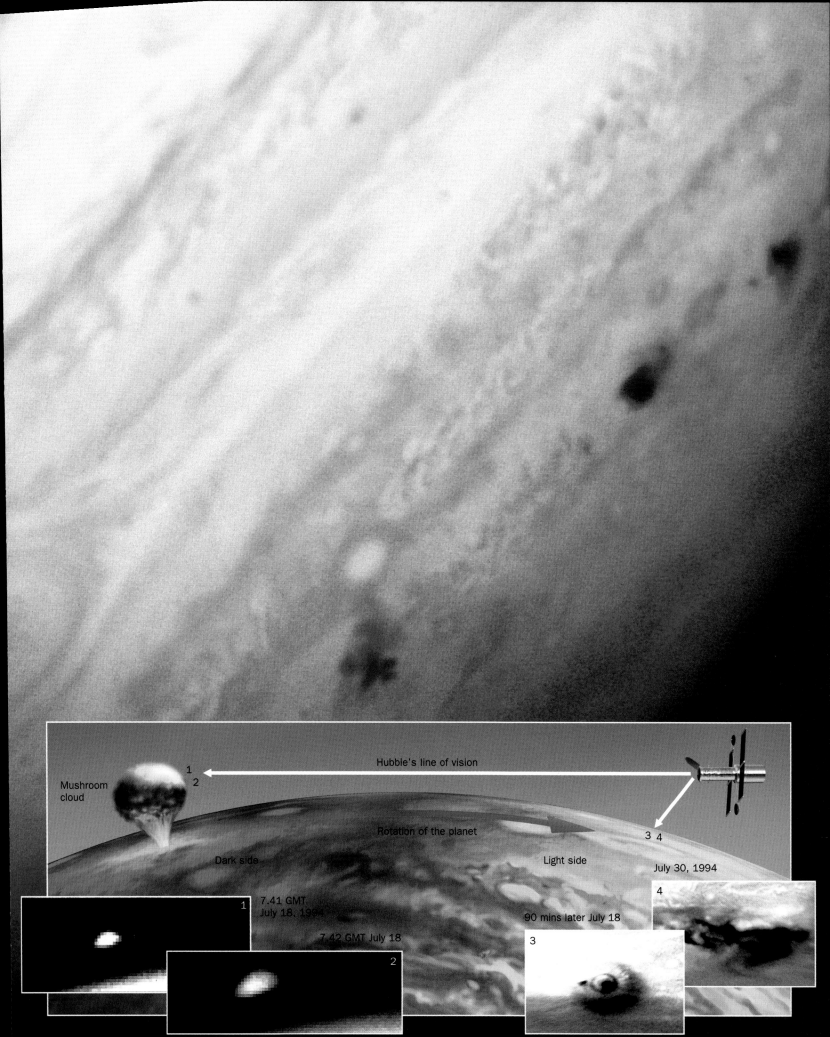

Hubble's line of vision

Mushroom
cloud

1
2

Rotation of the planet

3 4

Dark side

Light side

July 30, 1994

7.41 GMT
July 18, 1994

7.42 GMT July 18

90 mins later July 18

1

2

3

4

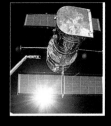

black, horseshoe features in the upper atmosphere of the planet. In the days, weeks and months that followed, the orbiting telescope recorded the way that the crash sites changed. The succession of images allowed astronomers to watch the action of the wind and weather patterns on the planet in precise detail.

As well as the small icy bodies in the outer Solar System, other small rocky bodies exist closer to the Sun. They are called asteroids and are predominantly found in the space between Mars and Jupiter. The asteroid belt, as it is called, is home to thousands of minor bodies ranging in size from Ceres (584 miles or 940 km across) to bodies less than 0.5 miles (1 km) across. Collisions between asteroids have knocked some out of the main belt. They revolve around the Sun in eccentric orbits and even cross the Earth's orbit in their passages through space. Telescopes around the world keep a constant watch for these Earth-crossing asteroids because of the danger they would present if they were to collide with us.

Collisions between asteroids have littered space with dust grains and pebbles. These are known as meteoroids, a term also used to describe the debris left behind by comets. Periodically, the Earth plows through a swarm of meteoroids and they are seen burning up in the atmosphere. The popular name for these incandescent sparks is shooting stars but they are correctly known as meteors. A meteor large enough to strike the Earth is called a meteorite. Meteorites have been found that appear to have originated on the Moon and Mars. In the past, very large meteorites must have struck these worlds, producing explosions that threw vast quantities of surface rock into space. Later, some of the debris collided with Earth, falling to the surface as meteorites. The largest meteorite to strike Earth in recent years, known as the Tunguska meteorite, fell over Siberia in 1908. It was probably 65-130 ft (20-40 meters) in diameter, exploded just above the ground and caused devastation over an area measured in hundreds of miles or kilometers. It was very fortunate that the meteorite did not fall over a populated area, otherwise the death toll would have been enormous.

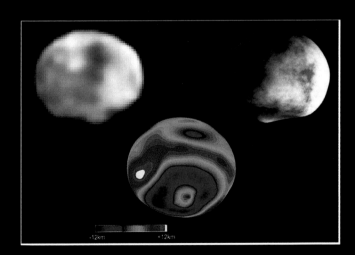

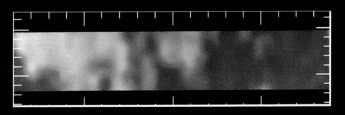

Objects in the Solar System orbiting the Sun *(right)*. A number of Earth-crossing asteroid orbits recorded in relation to the orbits of the planets. 1 Saturn, 2 Jupiter, 3 Earth, 4 Mars, 5 the main asteroid belt between Mars and Jupiter, 6 the Trojan asteroids, 7 Asteroid Hidalgo, 8 Asteroid 1983TB, 9 Asteroid Apollo, 10 Asteroid Icarus, 11 Asteroid Eros. The yellow lines show the calculated orbits of three meteorites that have struck the Earth. 12 Meteorite Pribram, 13 Meteorite Lost City, 14 Meteorite Innisfree.

Evidence of a giant impact on Asteroid Vesta *(left)*. The Hubble image (top left) shows a curious "nub" near the asteroid's south pole. The image (center) resulted from combining 78 images of Vesta as it rotated. This image has been tilted upward to show the 285 mile (460 km) wide impact basin (blue) as an oval with a central peak (orange and red). The computer generated model of the asteroid (top right) shows that the "nub" of the original image is the central peak of the impact basin.

Surface of Asteroid Vesta *(left)*. The colors indicate different rock compositions and types, from lava flows to impact basins.

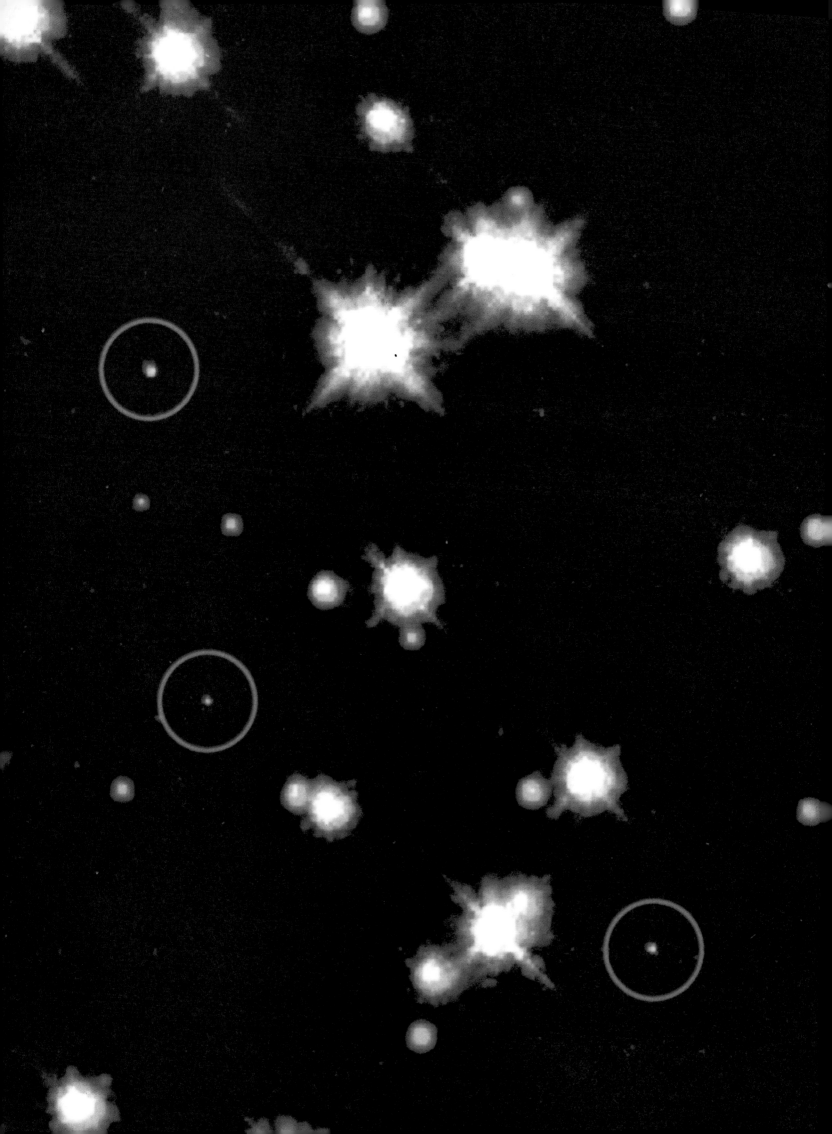

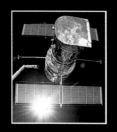

THE
STARS

STARS ARE CELESTIAL POWERHOUSES; they are born from dusty clouds and shine for millions and billions of years. Some are in our own galaxy, and others can be millions of light years away in other galaxies. The Hubble Space Telescope has allowed us to see unprecedented views of the different types of stars, and has helped astronomers to discover long sought-after failed stars, known as brown dwarfs.

THE VARIETY OF STARS is dazzling. The most noticeable features are the range of colors and the brightness of the light they radiate. Color can tell an astronomer a lot; it can be indicative of the size or the age of the star. Like people, stars pass through many stages during their lifetimes. They are born, enter an unstable adolescence, become adult, grow older, begin to function erratically and eventually die. The majority of stars are known as main sequence stars and can be thought of as analogous to human beings during their adult working lives.

Stars also come in a bewildering array of sizes. There are giant stars and dwarf stars, stars that contain a lot of mass and less massive stars. Sometimes they cluster in double, triple or even quadruple star systems where they orbit around one another, in the same way that the planets orbit the Sun in our own Solar System. As yet there has been no direct confirmation of planets orbiting other stars, although circumstantial evidence has been found.

White dwarfs *(right)*, ringed in blue, in the globular star cluster M4. They are the small, hot, compact stellar remnants left behind by some main sequence stars. Most of the stars around them here are red giants.

10,000

10

Brightness (Sun = 1)

1

0.1

White dwarfs

16,000   11,300   8(

The Hertzsprung-Russell diagram *(below)* is crucial to astrophysicists. Every type of star can be plotted on it. Surface temperature is indicated on the scale that runs along the bottom from left to right (cool stars on the right) and brightness on the scale from top to bottom (dim stars at the bottom). The curving diagonal feature (top left to bottom right) is known as the main sequence and is where the majority of stars are plotted.

Red giants, red dwarfs and brown dwarfs *(below)*. A red giant such as Betelgeuse (top) is so enormous that if it were to replace the Sun, it would engulf the inner Solar System. Gliese 623b (center, the object to the right of the main star) is a red dwarf. Failed stars, such as brown dwarf Gliese 229b (bottom, the tiny object center image) are so cool and dim that they are off the bottom scale of the diagram.

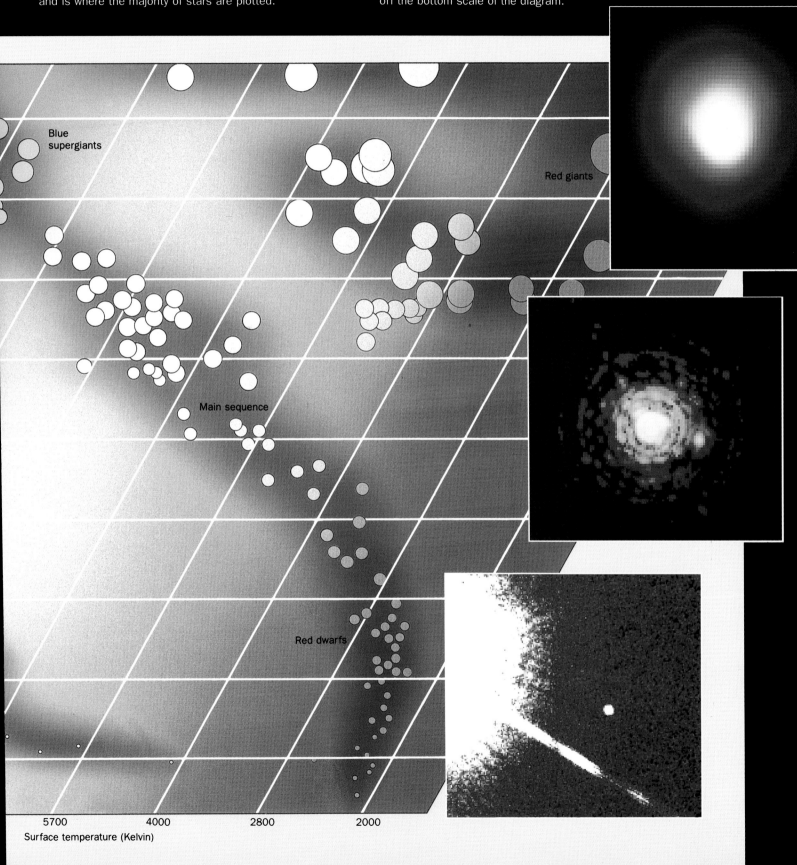

Blue supergiants

Red giants

Main sequence

Red dwarfs

5700   4000   2800   2000

Surface temperature (Kelvin)

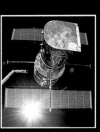

BIRTH OF STARS

STAR FORMATION is a lengthy process, which astronomers can observe taking place in many areas throughout our galaxy. Stars begin to form when regions within giant clouds of gas collapse under their own weight. A single cloud of gas, known as a giant molecular cloud, contains the raw material to make many thousands of stars. They do not all form at once, however. It takes millions of years for a molecular cloud to convert all its bulk into stars. During the course of this time, generations of stars live and die within the cloud's confines, and the way that they live and die affects how the next generation of stars is born. This is because, throughout their lives, stars give off radiation and eject tiny particles smaller than atoms. These emissions push the cloud material around, causing areas of greater pressure that increase the density of the cloud in patches, assisting star formation. Clumps begin to form in the densest regions, and at the center of a clump a protostar will emerge. A protostar is the central conglomeration of mass that will eventually become a star, once it has collected enough matter. A vast amount of pressure needs to build up in the central regions of the protostar in order to ignite the nuclear reactions that

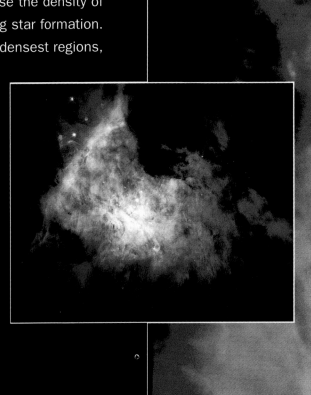

The Orion nebula *(bottom left, main image and above)* is 1,600 light years away from Earth and is a region of on-going star formation. There are many young stars in the region as well as protostars in formation. The different colors in the image are caused by the presence of different gases: red is nitrogen, blue is oxygen and green is hydrogen (mixed oxygen and hydrogen produce a turquoise blue). The whole central region of the nebula is pictured bottom left. Part of it has been enlarged (main image) to show Solar Systems in formation. They are the bright disks enlarged again (above). At the center of each disk is a protostar; the disks themselves are where planets are forming. The emerging Solar Systems are called protoplanetary disks (proplyds). This is the first time that astronomers have been able to observe them.

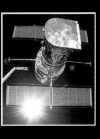

will make it shine. As more and more particles of gas crowd down onto the protostar, the pressure at the center rises dramatically and the temperature at its core becomes greater and greater until it triggers the process of nuclear fusion that will cause the star to give out energy for the rest of its life.

Stars are always composed of exactly the same material in exactly the same proportions. This is believed to hold true throughout the universe and the formula is known as the cosmic abundance of elements. Hydrogen gas makes up 75 percent of a star and helium gas constitutes another 23 percent. The remaining 2 percent is made up of other chemical elements, including gases, metals and radioactive substances. The determining factor that makes one star evolve differently from another is its mass. At any time in a star-forming region, the process of collapse will be happening in dozens or even hundreds of clumps simultaneously. Each of these clumps will contain a different amount of matter and consequently each star will be born with a different mass. In this way, a whole range of stars are born from a single cloud.

Stars with smaller masses tend to be more numerous than those with larger masses. A star like the Sun, for instance, which is known as a yellow dwarf, will be produced in much larger quantities than a blue supergiant. Red dwarfs, the smallest main sequence stars, will be the most numerous but they can be quite difficult to detect. Blue supergiants are much easier to see because they are so luminous. They are the most massive stars that we know of, sometimes containing 50 or even

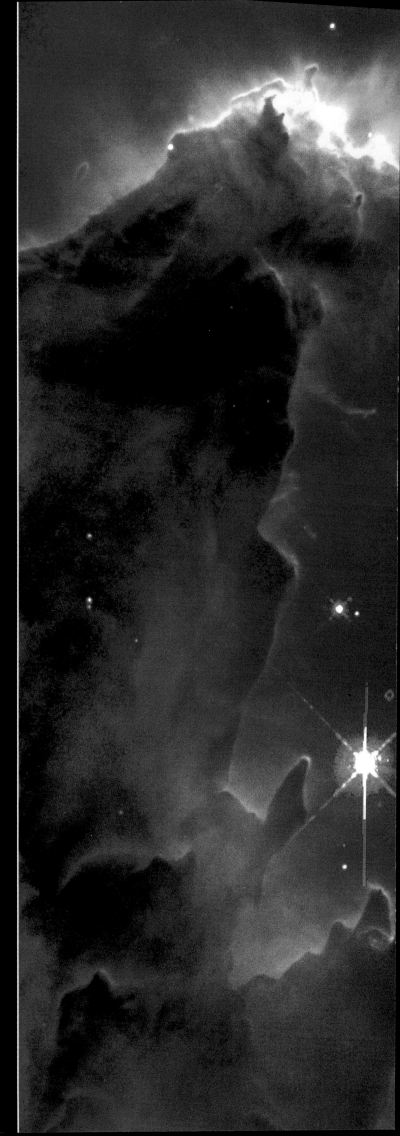

The Eagle nebula *(main image)* is another star-forming region. These vast pillars are not "key-holes" through the surrounding gas but columns of dusty material in which hundreds of stars are forming. The pillar on the left is one light year long; if the Solar System were placed at one end, the far end of the pillar would reach only a quarter of the way to our nearest star, Alpha Centauri. This suggests that stars must drift apart after they form.

A ground-based telescope's view *(below)* of the Eagle nebula. The pillars can be seen in the center of the image (the middle pillar is the most obvious of the three). Even at this distance, astronomers can identify them as absorption nebulae in front of an emission nebula. The red glow is characteristic of hydrogen gas, which is illuminated by blue supergiant stars. The emission nebula is only a small part of the overall molecular cloud.

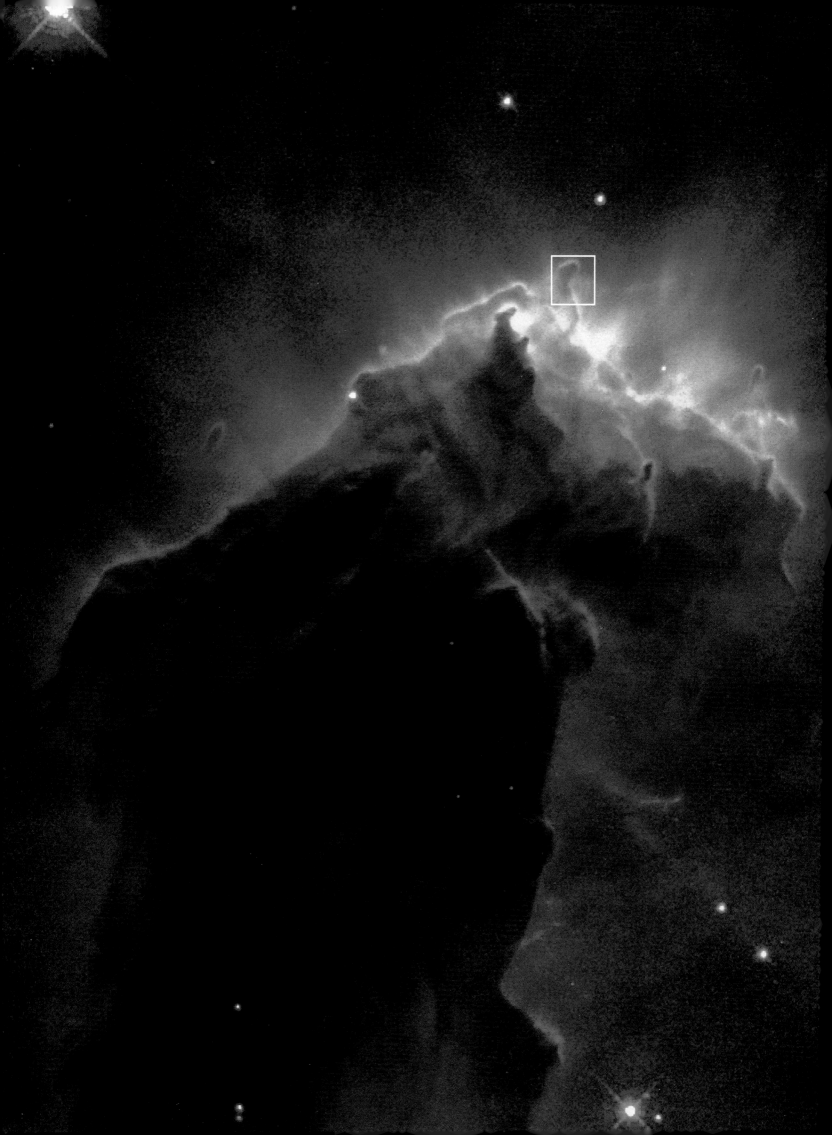

100 times the mass of the Sun. The pressure and temperature in these stars are so phenomenally high that they produce enough radiation to literally light up the gas in the cloud from which they were born. Blue stars give out huge quantities of ultraviolet radiation, which can then be converted into visible light by the gas atoms in the surrounding molecular cloud. This is why star-forming regions are almost always signposted by large patches of glowing hydrogen and other gases, known as emission nebulae. During its time in orbit, the Hubble Space Telescope has taken many images of them. If a nebula is glowing, it is because its constituent atoms have been excited by blue supergiants that formed in the previous million years or so – quite recently, on the astronomical scale. These glowing clouds of gas should not be confused with absorption nebulae, which are dark clouds of gas seen in front of stars or emission nebulae. The material within absorption nebulae is yet to be turned into stars. The potential lifespan of a star varies dramatically depending on its mass at birth. A high-mass star such as a blue supergiant will "burn" its fuel so quickly that it will only live for a few million years. A red dwarf, however, will perform the same process but in such a comparatively meager way that it will survive for billions of years.

Following the successful birth of the first generation of stars, the molecular cloud is compressed by the stellar winds issuing from the new-born stars. The winds are made up of very fast moving particles that push the surrounding gas and compress the nearest region of the molecular cloud. This triggers more star-formation and the whole complex process begins again.

The end of the largest pillar in the Eagle nebula *(left)*.
Ultraviolet radiation, created by a previous generation of massive stars, is eroding away the dust and gas from the end of the pillars. As it does so, it reveals clumps of material becoming protostars. Hubble gave astronomers their first view of these protostars, finger-shaped objects protruding upward. They were subsequently named EGGs (Evaporating Gaseous Globules). These protostars have been forcibly halted from collecting more mass because they have been separated from their nascent cloud. The formative processes within a typical EGG (marked with a white box) are outlined below.

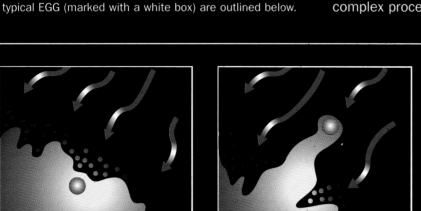
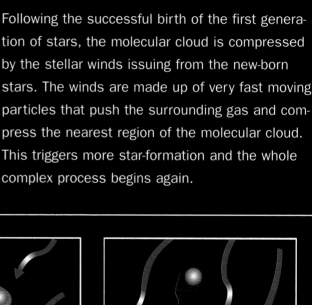
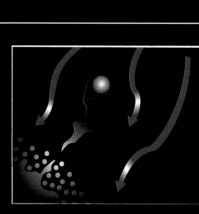

An EGG is formed when the ultraviolet radiation from a previous generation of massive stars illuminates the surface of the star-forming cloud, such as the pillars in the Eagle nebula, and begins to erode the dust. The surrounding gas is supplied with energy by the radiation and "evaporates" away.

As the cloud is eaten away, a clump of denser material is not so easily dispersed. This clump forms a tadpole-shaped object sticking out from the parent cloud. It is, in fact, a collapsing fragment of star-forming cloud, housing a protostar, which is still in the process of becoming fully fledged.

The EGG has been disconnected from the molecular cloud and will begin to evaporate. This will uncover a very young star, deep inside, which has previously been hidden from view. It is like the premature birth of a star that would have grown larger had it not been uncovered.

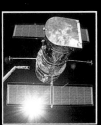

JETS AND DISKS

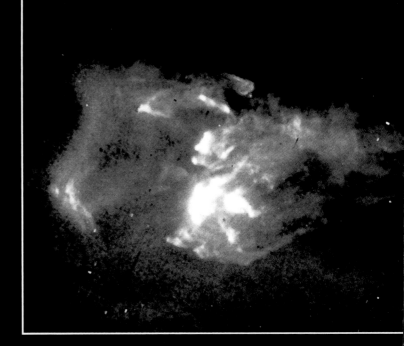

THE FORMATION OF STARS is not just a simple process in which matter condenses into clumps that, in turn, condense even further to become stars. The orbital motion of the giant molecular cloud around the center of our galaxy alters the trajectory of the in-falling matter so that it whirls toward the protostar, similar to the way that water spirals down a plug hole. The matter gradually creates a disk of material around the protostar's equator. In the dense environment of the disk, gas particles often collide and stick together to form dust. This is a process known as accretion and the disk is often referred to as an accretion disk or proplyd (short for protoplanetary disk). It is thought to be the first step in the formation of planets round the star because, after dust particles have formed, these too collide and produce bigger particles. The process continues until asteroid-sized objects, and then planets, have been made.

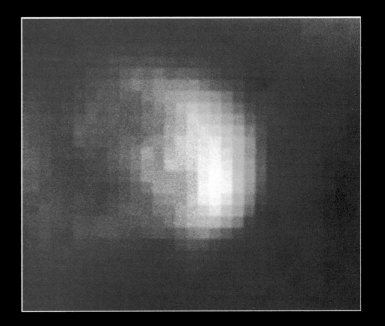

While all this is going on, the protostar itself is stirring into life. The first rumblings of nuclear fusion have begun deep inside and this is having a dramatic effect on the star and its immediate surroundings. The star generates a stellar wind, a constant stream of minuscule particles. These rush outward from the protostar and begin to blow away the in-falling dust. Around its equatorial regions, where the proplyd has built up, the density of matter is great enough to resist the stellar

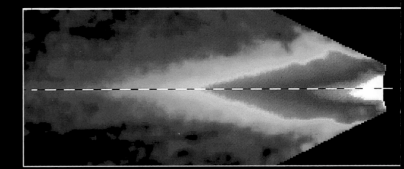

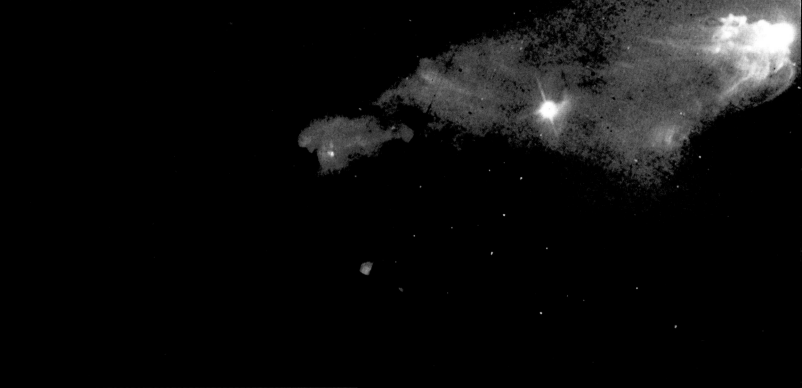

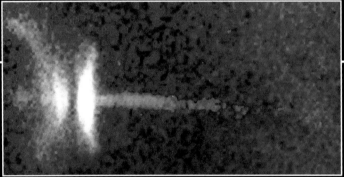

Jets from young stars *(top and inset).* Jets from the central star are difficult to see but the Herbig-Haro objects (top), created when the jets illuminate the surrounding gas, appear as orange plumes. The young star itself lies mid-way between the two objects but is too dim to be seen. The jets can be seen (inset) as can the proplyd, which is viewed "edge on" and appears as the black line running between the bright bowl-shaped features.

A protoplanetary disk from the Orion nebula *(left).* The proplyd is face-on which means that we are looking down on the "north pole" of the star. The orange point at the center of the disk is our view of the young star.

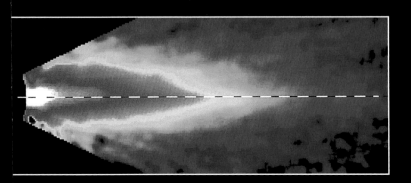

wind. At the rotation poles, however, there are relatively few dust particles and the stellar wind can punch holes through. The stellar wind appears as two jets, streaming away from the core star and illuminating the surrounding gaseous material. The "knots" of illuminated gas, found at the end of these jets, are called Herbig-Haro objects. In time, the stellar wind erodes away all the surrounding dust and gas. As the jets becomes wider they are known as outflows. Eventually, the stellar wind will dissipate the proplyd to reveal any planets that have formed there. Shortly afterward, the star begins to stabilize and becomes a main sequence star. It will remain stable and predictable until the supply of hydrogen in its core begins to run out.

A residual disk of material *(left)* found around the main sequence star Beta Pictoris and shown in false color to make the structure more apparent. To make the faint disk easier to see, the star in the center has been blotted out, and the disk begins 3.7 billion miles (6 billion km) from the center of the star.

STELLAR DEATH

THE STELLAR EQUIVALENT of retirement is entered when a star becomes a red giant. At this stage in its life, a great turning point has been reached; the hydrogen in its core has finally become too scarce to continue producing enough energy to keep the star stable. The core begins to shrink under the pull of gravity and its temperature and pressure soar. When the temperature reaches 100 million°C, the helium in the core is fused into carbon. This change in the nuclear reactions inside the star results in it becoming bloated in appearance. The helium reactions release so much energy that the outer layers of the star are pushed away from the core. The star swells up to many times its previous size, and because the outer layers have moved so far away from the core, they cool down and become red in color.

The Helix nebula *(top inset)*. The ejected gas from a dying star, which cannot be seen, has come into contact with surrounding gas in the interstellar medium. The force of the colliding gases has caused "cometary knots" to form. Cometary knots were named because of a passing resemblance to comets in the inner Solar System. However, each knot is about twice the size of the Solar System and the tails measure over one thousand times the distance from Earth to the Sun.

The Egg Nebula *(left)* is created by a dying star. The star is surrounded by a dense cocoon of dust, hiding it from direct view. Concentric shells of material have been thrown into space over the previous few hundred years and, where this material is thinner, starlight escapes in "search-light" beams. In the NICMOS image (bottom inset) dust appears blue, and hot hydrogen gas red. The movement of the gas carves the holes which allow the "search-lights" to shine through.

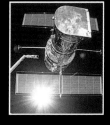

When the Sun reaches the red giant stage, some four billion years in the future, it will swell up so much that all of the planets from Mercury to Mars will be destroyed by its heat and probably swallowed up in its outer layers. Red giants are dotted all over the night sky and are often some of the brightest stars in the universe because they radiate so much energy from their enormous surfaces. After several million years of this existence, a red giant runs out of the helium that it needs to fuse into carbon. Once this happens, the star's days are numbered, but the manner in which in dies depends upon how much mass the star possesses.

One possibility is known as low-mass stellar death and occurs in any star with a mass up to five times that of the Sun. The second possibility is known as high-mass stellar death, when the star contains more than five times the mass than the Sun. In this case it suffers the spectacular fate of a supernova. During low-mass stellar death, the death throes are muted and almost peaceful. The core will contract once more but it will never reach the stage where carbon fusion can take place. Instead, only a shell around the core will ignite and begin to fuse helium into carbon. As a result

of this reaction, the star will become unstable and its surface will begin to pulsate. For a time, it will become a variable star, constantly changing its brightness and size. During these oscillations it will eventually push its outer layers right off into space. These layers will drift away from the star, creating what is known as a planetary nebula. This name is very misleading because the nebulae produced by this process have nothing whatsoever to do with planets. The term was first coined by scholars of antiquity, who could only resolve the discarded layers as fuzzy disks in their telescopes. Superficially, they bore a resemblance to the planets and so were named after them. As the star sheds up to to 80 percent of its mass in this way, the inert core is finally exposed. By this stage it will have collapsed into an incredibly dense stellar remnant known as a white dwarf. These little stars are smaller than Uranus, but may have about the same mass as the Sun.

Planetary nebula NGC 7027 *(right)*. The tenuous outer atmosphere of a red giant is thought to present exactly the right conditions for the formation of dust. In this case, dusty patches show up as a network of red clouds. They are surrounded by concentric blue shells of gas, which are the ejected outer layers of the dying star. The creation of a planetary nebula such as this usually takes a few thousand years. The emerging white dwarf is visible at the center (see arrow). The colors in this image are not real, but chosen to aid visibility of the features.

This beautiful planetary nebular *(left)* is known as M2-9. What looks like a single, central star is actually a pair of incredibly close stars. They are so near they might even be touching. As they slide past one another, they throw off gas that forms a disk of material in their orbits. This disk deflects other material upward and downward, creating the symmetrical lobes of glowing gas which can be seen here. The glowing gas in this image is oxygen and nitrogen.

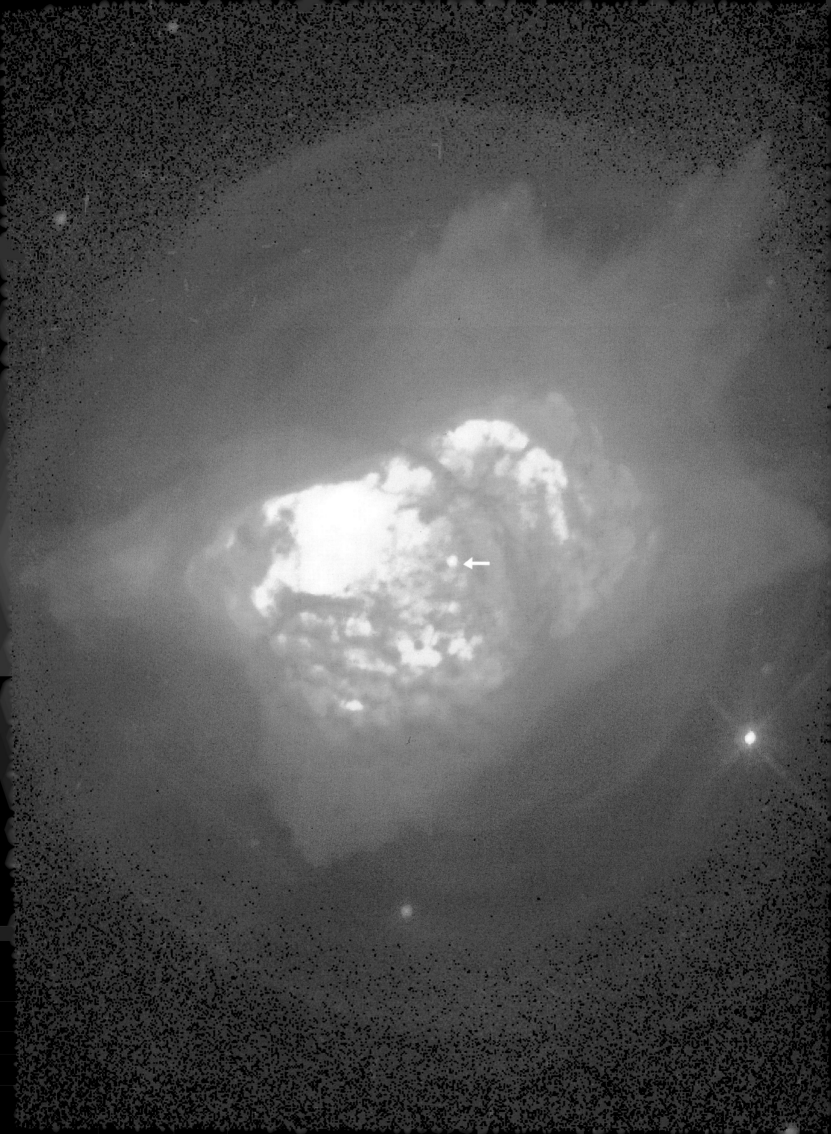

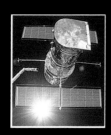

SUPERNOVAE

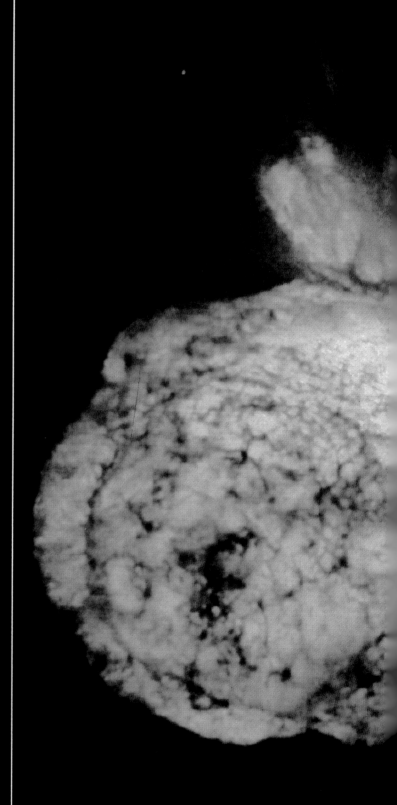

STARS THAT HAVE MORE than five times the mass of the Sun are known as high-mass stars. They die in spectacular explosions called supernovae. In the early stages of death (until they reach the end of the helium-fusing era) they evolve in a way broadly similar to low-mass stars. At this point the nuclear reactions in a low-mass star cease and a planetary nebula is formed. In a high-mass star, the weight of material on the inert core is so great that it can exert enough pressure for carbon fusion to take place. When carbon fuses it produces elements such as oxygen and neon, which in turn will eventually be fused to produce sodium, magnesium, silicon and sulfur. The final type of nuclear fusion that takes place in a high-mass star converts the previously synthesized chemical elements into iron, nickel and cobalt. All the elements synthesized in the star up to this point have actually given out energy during the fusion process. It is this energy, known as the binding energy, which has kept the star shining for millions of years.

Iron, however, is so stable that it requires quite a large input of energy before it will fuse

The supermassive star Eta Carinae *(right).* The blue supergiant, which can be glimpsed as the blue-white region in the center of the image, is estimated to be about 100 times the mass of the Sun. About 150 years ago it erupted violently making it, temporarily, the brightest star in the southern sky. The after-effects of this eruption are the dumbbell-shaped lobes of dust and gas which can be seen surrounding the star. Eta Carinae is doomed to become a supernova.

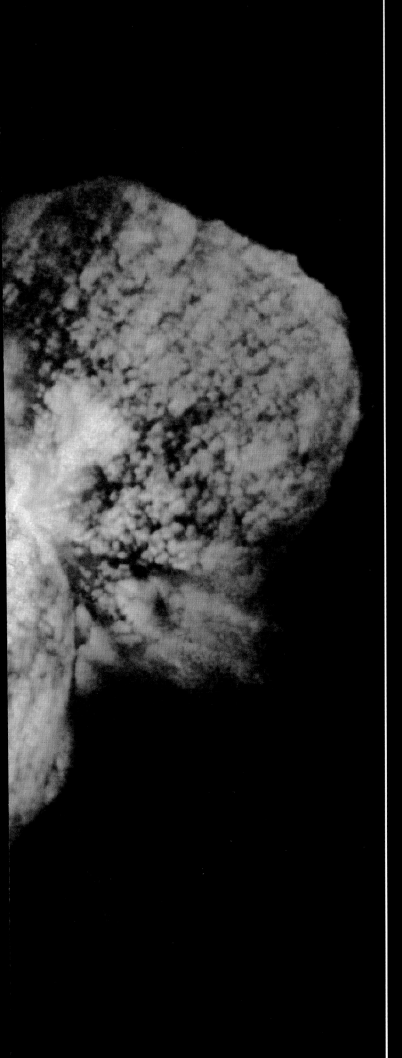

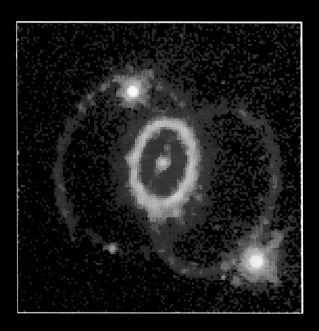

SN1987A *(above)* provided astronomers with their best chance so far to study a supernova at close quarters. Hubble revealed three remarkable rings surrounding the exploded star. The brightest is a "light echo" of the blast, bouncing off the surrounding dust and into our line of sight. The others are more mysterious; they may be caused by high speed jets of ejected particles.

together. The build-up of iron in the core is a watershed in the death of the star. The iron accumulates like ash in a log fire, but is totally inert until it achieves just under one-and-a-half times the mass of the Sun. At this point, the atomic structure of the iron spontaneously changes. Until this time, the iron atoms have been strong enough to support their own weight. Having achieved this mass, however, the electrons that orbit the atomic nuclei are forced out of orbit and merge with the nuclei. This significantly reduces the volume of space that the atoms occupy, and the core collapses, shrinking to a tiny fraction of its former size in less than a second. With nothing left to support the rest of the star, it too begins to collapse. Matter rains onto the collapsed core producing tremendous impacts of gaseous material. These collisions initiate a number of nuclear reactions, so many that

they produce a shockwave that disrupts the collapsing star, producing a supernova explosion. A supernova is so powerful that for a few weeks it outshines all the other stars in the host galaxy put together! In this incredible outpouring of energy, all the elements heavier than iron are synthesized. There are no other mechanisms known that are capable of building these heavy elements and so we are forced to conclude that most of the atoms in the Earth and even in our bodies were once at the center of a massive star.

Just as a low-mass stellar death leaves behind a white dwarf and a planetary nebula, so a high-mass stellar death results in two new celestial objects, a supernova remnant and a stellar remnant. A supernova remnant is the gaseous nebula produced by the explosion. During the very early stages of its life, it will be a compact object surrounding whatever is left of the central star. Even after the initial fury of the explosion has died, the

remnant will glow brilliantly due to the radioactive decay of elements forged during the supernova. Later, the supernova remnant can sometimes derive power from the stellar remnant left at the center of the cloud. Planetary nebulae drift away from their central stars, but supernovae remnants are expelled at great velocity and expand through space. The Hubble Space Telescope has such sensitive optics that it can trace the changes in these expanding clouds of gas. Over the course of time, they expand outward and fade. If they occur in giant molecular clouds or particularly gaseous regions of space, the expanding supernova remnant will collide with the slower moving gas and create shock waves that emit light. Very often, these tell-tale filaments of emission are the only

The Crab nebula *(right)* is the supernova remnant of a star that exploded in 1054. The inset (top) shows the best ground-based image of the region. Hubble has focused on the pulsar at the center of the nebula (main image). This tiny object spins 30 times every second and bathes the nebula in particles that have been accelerated in the pulsar's magnetic field. This provides the energy to make the nebula glow.

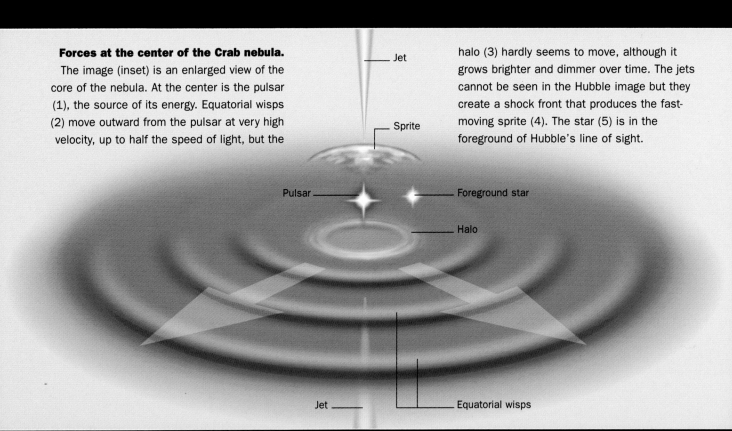

Forces at the center of the Crab nebula.
The image (inset) is an enlarged view of the core of the nebula. At the center is the pulsar (1), the source of its energy. Equatorial wisps (2) move outward from the pulsar at very high velocity, up to half the speed of light, but the halo (3) hardly seems to move, although it grows brighter and dimmer over time. The jets cannot be seen in the Hubble image but they create a shock front that produces the fast-moving sprite (4). The star (5) is in the foreground of Hubble's line of sight.

Jet

Sprite

Pulsar

Foreground star

Halo

Jet

Equatorial wisps

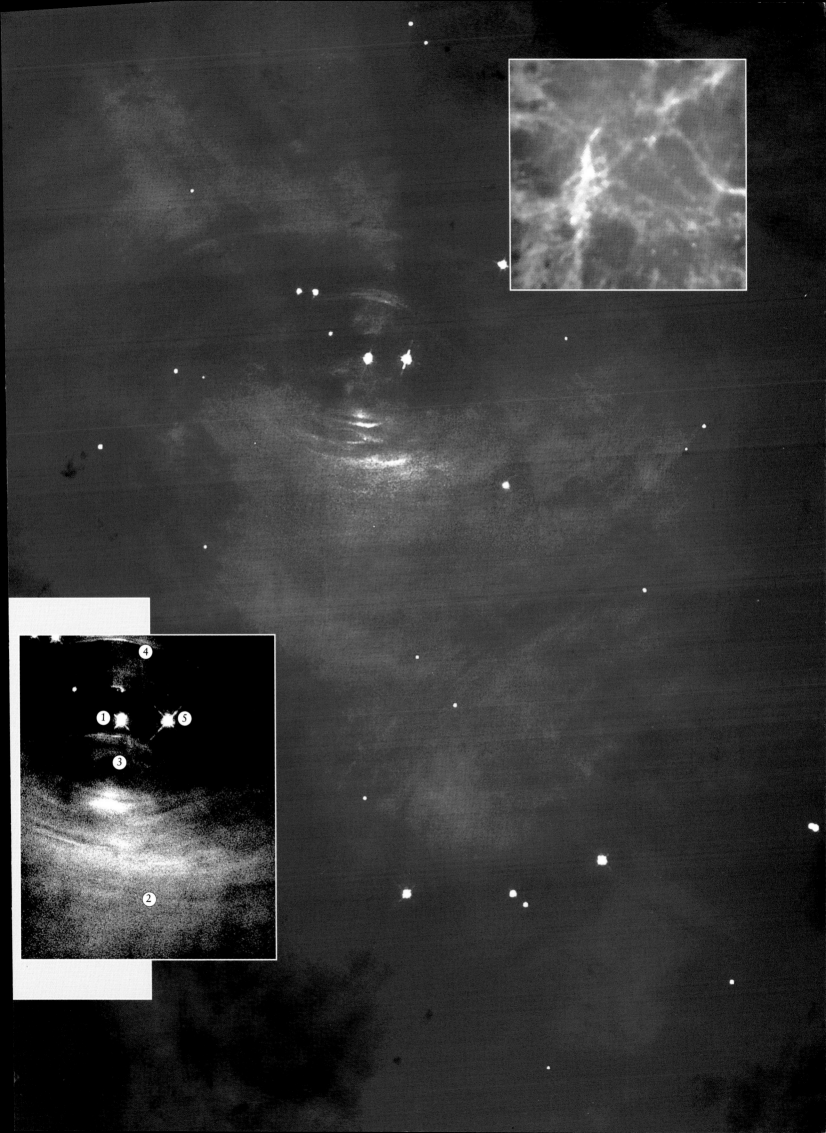

indication of older supernova remnants. The stellar remnant may become either one of two things, depending on its eventual mass. When the iron core collapses, it is pummeled by more stellar material that rains down on its surface. The eventual mass of the core is always greater than one-and-a-half times the mass of the Sun. If it is also less than three-and-a-half times the mass of the Sun it will become a neutron star. This is an intriguing object that squeezes its mass into a sphere of only 9–12 miles (15–20 kilometers) in diameter. If the collapsed core contains more than three-and-a-half times the mass of the Sun, it will become a black hole. These are truly bizarre destructive forces, sucking out of existence anything that strays too close. Technically speaking, there is a second type of supernova. It can occur in a binary star system containing a red giant and a white dwarf. The red giant swells up so much that its outer layers enter the gravitational influence of the white dwarf. This causes material to be funneled from the red giant onto the white dwarf. The pressure this exerts on the interior of the white dwarf is often sufficient to begin a runaway nuclear reaction, which totally destroys the white dwarf.

Sections of the Cygnus Loop (*main image and inset*). The Loop is a bubble-shaped shock wave, which has been expanding since a supernova explosion 15,000 years ago. The wave is caused by the expanding supernova remnant compressing and heating the gas in its surroundings. The blue streak in the main image is a knot of gas which was also expelled by the supernova. It is like an interstellar "bullet" travelling through space at fantastic speeds. In this false-color image, oxygen has been coded blue, hydrogen green and sulfur red. Compression waves like this one can often precipitate the formation of new stars. Scientists believe that our own Sun was formed in this way.

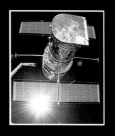

THE
GALAXIES

*T*HE STARS THAT WE CAN SEE IN THE NIGHT SKY *are all contained within our own galaxy, the Milky Way. Astronomers estimate that the Milky Way contains billions of stars. Using telescopes to look deeper into space, we can also see other galaxies. These stellar cities also contain billions of stars and are many millions of light years away. Some appear to be similar to the Milky Way, while others are very different.*

THE MILKY WAY

O N A CLEAR NIGHT, well away from the blinding light pollution of major cities, the night sky appears to be bedecked with jewels. Approximately 3,000 stars are visible to the naked eye. As well as the individual stars, a misty band of light stretches across the sky. This is our view of the Milky Way and is caused by the shape of our galaxy and where we are positioned within it. When viewed through a telescope, the Milky Way is seen to be composed of millions of individual stars. The shape of our galaxy proved an intriguing problem to astronomers for many centuries. Some thought that the Sun was located in a more or less spherical conglomeration of stars. Others recognized that the band of light is our view (from the outer edges of the spiral) of a disk-like structure that constitutes the major part of the galaxy. It is now known that our galaxy has a central bulge of stars, which is surrounded by a relatively thin spiral pattern of other stars. The Solar System is located in one of these spiral arms and what we see in the night sky is our view of the central bulge and some of the other arms. The furthest arms appear to be superimposed upon the closer ones. Superimposition occurs when a distant object appears to be side-by-side with a much closer one. Galaxies and other celestial objects are often referred to as being located in specific constellations. This is just to aid their identification, and does not mean that the object is located at comparable distances to the stars that comprise the constellation.

A typical spiral galaxy (right). Hubble cannot take pictures of our galaxy, the Milky Way, because the Space Telescope is in orbit inside our own Solar System, but images of other spiral galaxies allow us to see what ours looks like. If this were our galaxy, the Solar System would be at position 1. Some other prominent features would be: 2 The Perseus arm, 3 The galactic center, 4 The Cygnus arm, 5 The Centaurus arm.

Our view of the Milky Way (below) is very different, because our viewing point on Earth corresponds to position 1 in the photograph. Seen with the naked eye the Milky Way appears to us as a fuzzy band of stars. The galactic center, Cygnus arm and Centaurus arm would be seen across the night sky at 3, 4 and 5 respectively.

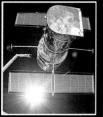

STAR CLUSTERS

I
N OUR GALAXY, AND THROUGHOUT THE UNIVERSE, stars are born in groups in dusty regions. Some spend a few hundred thousand years in clusters before drifting apart to become lone wanderers in their own galaxy. The clusters containing stars that drift apart are known as open clusters. They are found dotted randomly throughout the galaxies. Once their component stars have scattered, it is almost impossible to identify which were formed in the same cluster. Other stars spend their entire lives in the clusters where they were formed. These are known as globular star clusters, because they form globe-shaped masses.

A globular cluster orbits the center of the host galaxy. Usually this means that it is far away from the galaxy's main stellar region, although it will plunge through the galaxy twice in its orbit. Globular clusters typically contain 100,000 to 10 million stars packed into a spherical region of space only some 100 light years in diameter. They are almost always old structures and contain some of the most ancient stars in the universe. Experts believe that most globular clusters formed shortly after the birth of the universe when their host galaxies were still accumulating the material to begin star formation in earnest. Globular clusters around the Milky Way have been quite well studied. In fact, it was their positioning in the sky that first alerted astronomers to the idea that the Solar System could not be at the center of our own galaxy. In the 1920s, US astronomer Harlow Shapley noticed that there are a great many globular clusters in the southern constellations and far fewer in the northern constellations. Assuming they were evenly distributed, he concluded that we are on the outskirts of our galaxy, with the center in the direction of the constellation Sagittarius. His theory has been proved using radio telescopes. Hubble has made high-resolution study of globular clusters in distant galaxies a viable exercise. This work allows us to compare clusters in our own galaxy with those much farther away which, in turn, helps to determine whether we live in a typical galaxy. So far, it seems as if we do.

Globular cluster G1 *(right)* orbits the Andromeda galaxy which is the nearest large spiral galaxy to the Milky Way. This Hubble image is of comparable quality to ground-based shots of globular clusters surrounding our own galaxy, even though G1 is nearly 100 times farther away. The orange dots scattered across the image are red giants, and the two large, bright objects are foreground stars in the Milky Way. Hubble is studying some 20 globular clusters around Andromeda.

The center of distant galaxy NGC 1275 *(below)* is surrounded by approximately 50 massive, globular star clusters. The central region of this galaxy appears to be blue, and dark lanes of dust crisscross the image. The globular clusters show up as tiny brilliant blue points of light, which is surprising because it suggests they were recently formed, whereas most globular clusters are old.

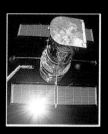

TYPES OF GALAXIES

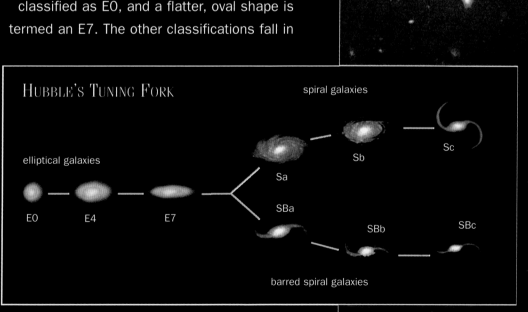

DEEP IN SPACE, far beyond the stars in our own galaxy, astronomers can see other galaxies, each teeming with billions of stars. Some galaxies are spiral in shape, similar to the one that we live in, others are elliptical, some defy description. The first scientist to make a systematic study of the various galaxy shapes was Edwin Hubble in the 1920s. As part of his PhD thesis, Hubble noted that galaxies were usually either spiral or elliptical. He went on to observe that spiral galaxies sometimes had a bar-like structure of stars connecting the nuclear bulge to the spiral arms. Based on his study of galaxies, Hubble devised a classification scheme that is still in widespread use today. The classification scheme subdivides elliptical galaxies into eight different categories, depending on how flattened the galaxy appears to be. A spherical galaxy is classified as E0, and a flatter, oval shape is termed an E7. The other classifications fall in

HUBBLE'S TUNING FORK

spiral galaxies

elliptical galaxies

Sa

Sb

Sc

E0   E4   E7

SBa

SBb   SBc

barred spiral galaxies

between these two extremes. Spirals are divided into six subgroups. This time, the determining features are the size of the nucleus and the proximity of the spiral arms. If a galaxy has a large nucleus which has spiral arms wrapped tightly around it, it is classified as Sa. If it displays a bar of stars across its nucleus as well, the classification is changed to SBa. Sc or SBc galaxies are those that display small nuclei and loosely wound spiral arms. Our galaxy is usually referred to as an Sb type, although some astronomers believe that it may have a small bar which would make it an SBb. There are some difficulties inherent in classifying galaxies solely by appearance. One problem is that we cannot be sure which angle we are viewing from. When a spiral galaxy is seen from above, we can easily make out its beautiful arms; when it is edge on, the arms appear to

A field of distant galaxies *(main image)* contains some of the galaxy types classified by Edwin Hubble. NGC 4881, the bright shape to the left of the image, is an elliptical galaxy in a cluster of galaxies named after the constellation Coma. The spiral galaxy to the right of the image is also a member of the Coma cluster. Few of the other galaxies can be classified with any certainty because they are much farther away, part of an even more remote cluster. This image has been used to try to identify globular clusters around NGC 4881 in order to help astronomers calculate the distance from Earth to the Coma cluster.

Galaxy NGC 4639 *(inset, top)* is a good example of a barred spiral. On the Hubble tuning fork diagram it would probably be classified as an SBa.

The Hubble tuning fork diagram *(far left)*, named because of its shape, shows the various types of normal galaxy that populate the universe. Edwin Hubble assumed that galaxies formed as ellipticals and then gradually evolved into spiral and barred-spiral varieties. Astronomers now know that this is not true. In fact, it may be that elliptical galaxies are formed when spiral galaxies collide with one another.

be a thin disk with the nucleus bulging in the center. This makes its classification more difficult but not impossible. When trying to classify elliptical galaxies, astronomers must be very careful to ensure that, if a galaxy is classified E0, it is genuinely spherical and not, for example, an oval E7 galaxy viewed from one or other of the tapering ends.

In general, star formation only occurs in spiral galaxies, and then only in the arms, which explains why the arms of many spiral galaxies appear blue. Young blue supergiants have very short lives, only a few tens of millions of years, but they blaze with brilliant intensity. They light up the spiral arms where they form, but turn into supernovae long before enough of them can be created to illuminate the galactic disk. By contrast, elliptical galaxies are composed of older stars and have no sites of active star formation. Some galaxies do not fit into Edwin Hubble's classification scheme, and are known as irregular galaxies. Even irregular galaxies, however, can be subclassified. A type 1 irregular displays some vestige of spiral shape but a type 2 is just an undefined mass of stars. Lastly,

there are peculiar galaxies. These look normal but have some abnormal features that require additional explanation. Sometimes the peculiarity is a truly intense bout of star formation, a phenomenon known as a starburst. Galaxies vary enormously in size. A typical spiral galaxy such as the Milky Way is about 100,000 light years across. An elliptical may measure this in diameter but its volume would be much greater. Galaxies appear to be gregarious because they are almost never found in isolation. They gather in groups of a few dozen, or clusters of a few hundred. The Milky Way is part of a small group, known as the Local Group, which contains over 20 galaxies. The nearest large cluster is the Virgo cluster which contains hundreds of galaxies. All the galaxies in these associations are held together by mutual forces of gravity. Sometimes, a giant elliptical galaxy will dominate the center of a large cluster. Even clusters and groups can attract each other. These associations are known as superclusters and they form the largest known structures in the universe. Superclusters surround large voids in the universe. In fact, the structure of the universe is rather like Swiss cheese; the holes represent the voids, and the cheese is where the superclusters reside.

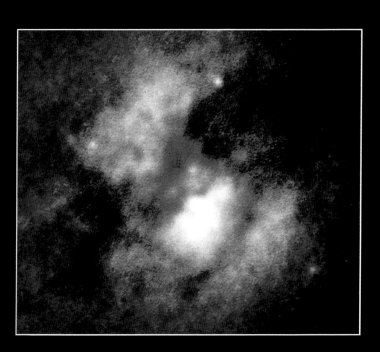

The large star-forming region NGC 604 *(right)* is located in the spiral galaxy M33, some 2.7 million light years away from Earth. At the heart of NGC 604 is a gaseous cradle containing over 200 hot newborn stars. The intense ultraviolet radiation from these stars causes the surrounding gas to glow. NGC 604 is on the leading edge of one of M33's spiral arms; this is exactly where astronomers would expect to see star-forming activity.

The peculiar galaxy Arp 220 *(left)* appears to be abnormally bright when viewed with infrared cameras. This Hubble Space Telescope image of the galaxy's core explains why excess emission occurs. It reveals gigantic star clusters, ten times larger than any previously known, at the center of the galaxy. They form a starburst that produces massive radiation. Some astronomers believe that starburst activity is caused when two galaxies merge, compressing their respective star-forming clouds into an accelerated phase of activity.

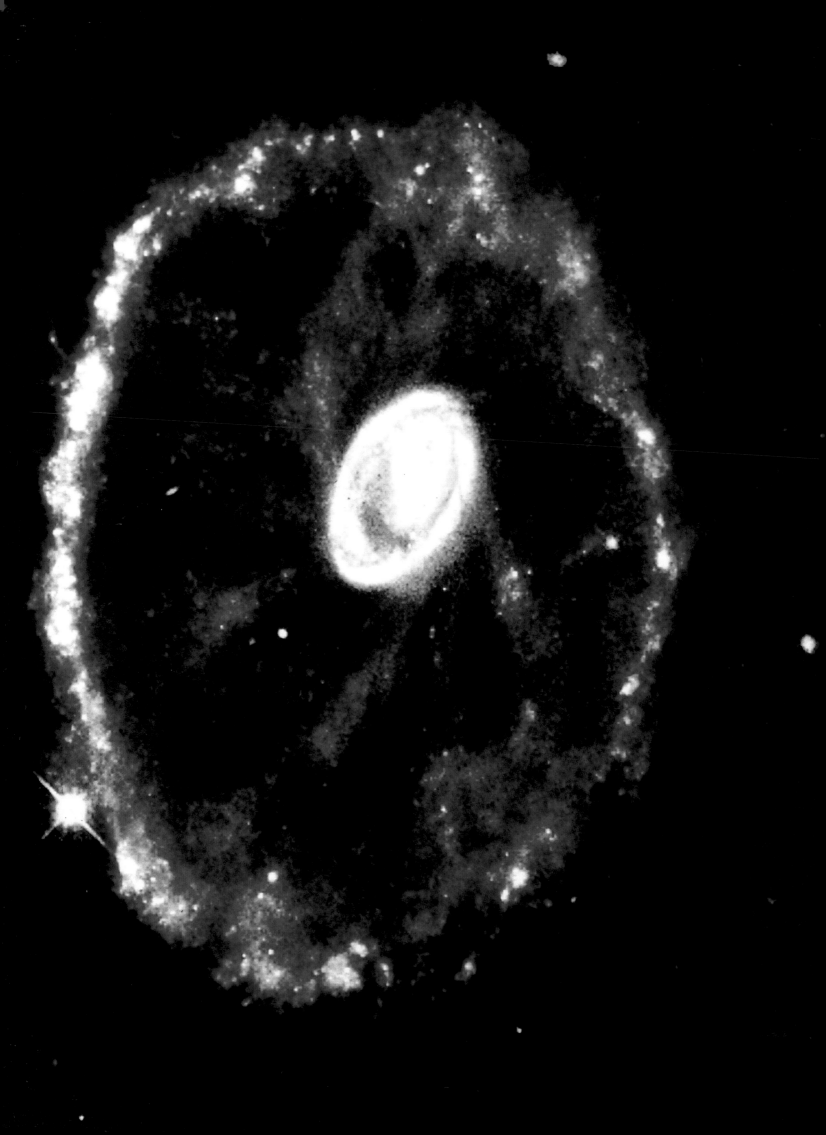

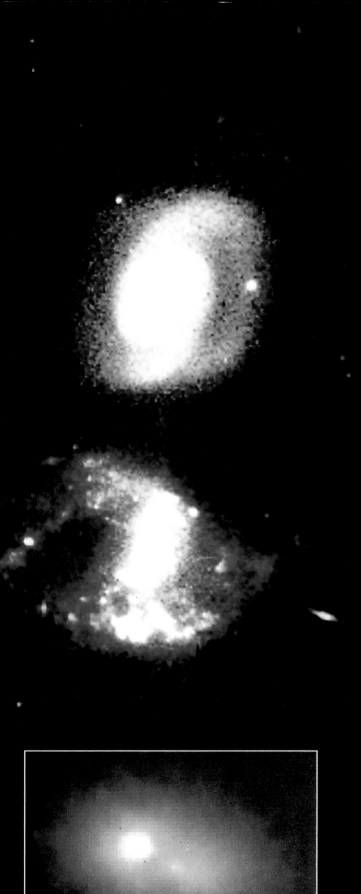

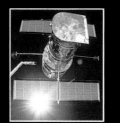

COLLIDING GALAXIES

A GALAXY IS IN A PERPETUAL STATE OF MOTION as it orbits the center of the cluster of galaxies in which it exists. As galaxies follow their respective orbits, they may, periodically, collide with one another. Because they are so vast, it takes thousands or even millions of years for the collision to run its course. Astronomers see the interaction as if frozen, at a specific moment in time. If the colliding galaxies are very different in size, the larger one will be relatively undisrupted, but the smaller galaxy will be completely devoured, losing its identity. If the colliding galaxies are comparable in size, both will be distorted, perhaps beyond all recognition compared with their former appearances. A collision between galaxies is believed to trigger massive bursts of star formation. The shock waves caused by the collision ripple through the gaseous material, compressing the interstellar medium and boosting the first stages of stellar formation.

This ring-like galaxy *(main image)* was once a spiral galaxy, through which a smaller galaxy (possibly one on the right of the image) has plunged during the previous few million years. The impact has caused a ripple, the edge of which continues to move outward from the center, marked by the young blue stars that it has caused to form. Just visible between the core and the ring are vestiges of its once prominent spiral arms.

The center of the Andromeda galaxy *(inset)* is revealed by the Hubble Space Telescope to contain two nuclei of stars. The main nucleus is the actual center of the galaxy, and the smaller nucleus is the remains of the core of a much smaller galaxy that is being consumed by Andromeda.

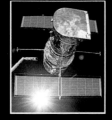

ACTIVE GALACTIC NUCLEI

ONE IN EVERY TEN GALAXIES contains what is known as an active galactic nucleus. This means that some extraordinary activity is taking place within the core, which has nothing to do with star formation or normal stellar processes. Any type of galaxy, spiral, barred-spiral, elliptical or peculiar, can possess an active galactic nucleus. If it does, it is usually referred to as an active galaxy. The energy generated by some active galactic nuclei is so prodigious that they totally outshine the stellar content of their host galaxies. The first active galaxies to be identified were the type in which the host was either a spiral or a barred-spiral galaxy. Examples of this type have now become known as Seyfert galaxies, after their

discoverer, Carl Seyfert, who began to observe them in 1943. They are distinguished by the fact that they possess overly bright nuclei, which (examination has revealed) cannot be produced by stars. A second type of active galaxy was found and eventually identified as being elliptical. These were named radio galaxies because of their extraordinarily large emission of radio waves. Radio galaxies were discovered by an amateur American astronomer named Grote Reber. Around the time that Seyfert was investigating active galaxies,

The center of radio galaxy 3C265 *(left)*. The blue lines (added by NASA) correspond to the direction in which jets of radio-emitting material are being squirted from the center of this active galaxy. The core itself is the bright white object in the center of the image. Surrounding it are a number of very large star clusters, or possibly very small galaxies. Astronomers detect that these clusters are being pulled into the core, possibly by the gravitational force of a black hole.

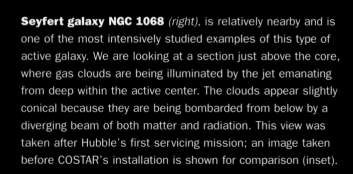

Seyfert galaxy NGC 1068 *(right)*, is relatively nearby and is one of the most intensively studied examples of this type of active galaxy. We are looking at a section just above the core, where gas clouds are being illuminated by the jet emanating from deep within the active center. The clouds appear slightly conical because they are being bombarded from below by a diverging beam of both matter and radiation. This view was taken after Hubble's first servicing mission; an image taken before COSTAR's installation is shown for comparison (inset).

Quasar QSO 1229–204 *(left)*. Usually quasars look just like stars but, if they are very powerful active galaxies, the Hubble Space Telescope should be able to resolve the surrounding galaxy. This image is one of the first ever to show evidence of the host galaxy in which the quasar is embedded.

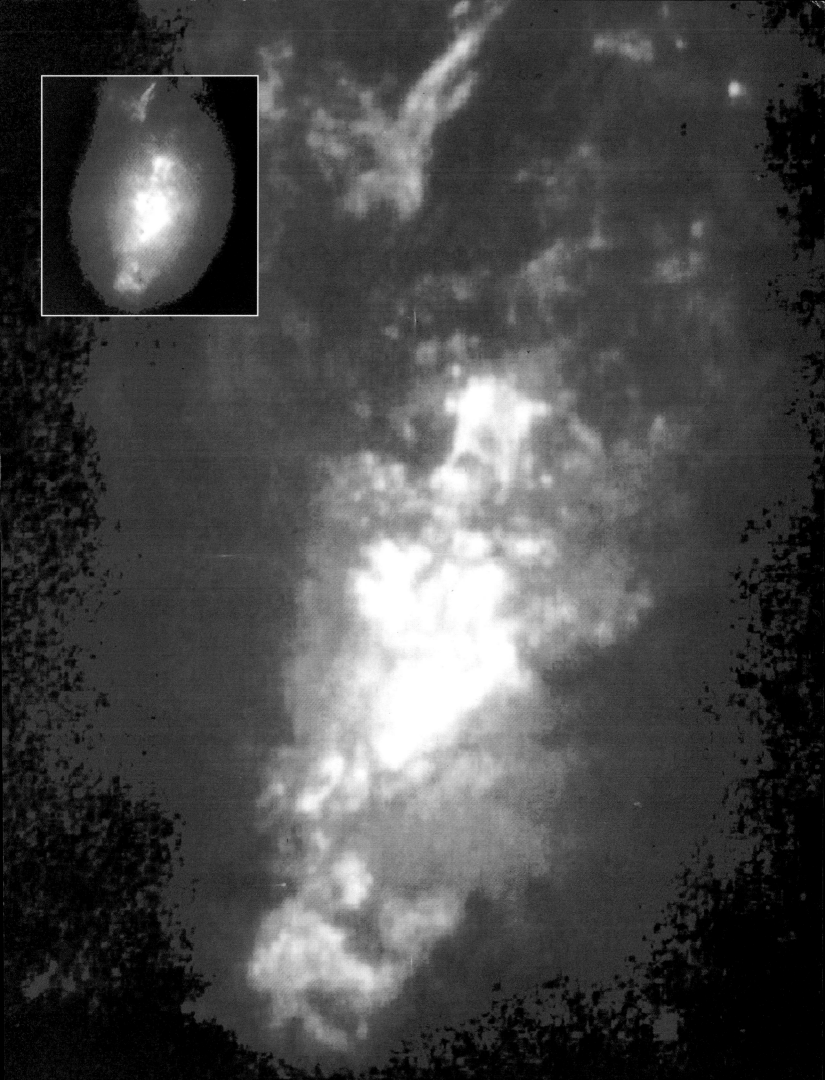

Reber was using a home-made radio receiver to map the night sky. His most interesting discovery was that the universe was "noisier" in some directions than others. One area that proved an extremely powerful source of radio waves was a small region inside the constellation Cygnus. Cygnus A, as that region is now known, was the first radio galaxy to be recognized by the scientific community. The radio waves emanate from gigantic lobes on either side of the host galaxy; they look as if they have been squirted out as jets originating from the galaxy's core. In some ways, the radio lobes resemble the Herbig-Haro objects created by the jets and disks around young stars, though the scale of the two structures is remarkably different. For example, the distance between two radio lobes can be millions of times the distance between two Herbig-Haro objects.

can be seen of the rest of the galaxy around a quasar; the active galactic nuclei is so incredibly bright that it blinds our view. In some quasars the radiation output varies by large amounts over short periods of time. Until their special status was recognized, these quasars were misclassified as variable stars. They are now known as blazars.

It was not until 1960 that experts realized that quasars were very distant objects indeed. A correlation of similar features between the various different types of active galaxies has allowed astronomers to build up a "working model" of the forces that power all these celestial objects. The area from which energy is generated is very small in relation to the amount of power that it produces. Measurements show it to be no greater than the diameter of our Solar System and yet the quasar outshines the combined brightness of billions of stars in the galaxy. The best theory to account for this phenomenon postulates a black hole at the center of an active galaxy. Black holes can be formed following the explosion of a supernova or in the center of a galaxy,

A third type of active galaxy is known as a quasar. The name is shortened from "quasi-stellar object" which reflects the fact that in a normal photographic image, a quasar looks virtually identical to a star. Only with careful analysis has it been possible to show that quasars are not stars in our galaxy, but incredibly distant celestial objects, which emit some of the most intense radiation in the universe. The radiation from a quasar bears a marked resemblance to that from Seyfert and radio galaxies, and this has led to the hypothesis that a quasar is also an active galaxy, but one in which we can only see the nucleus. Virtually nothing

The giant elliptical galaxy NGC 4261 *(left and top right)* is one of the brightest galaxies in the Virgo Cluster; it also happens to be a radio galaxy. The composite false-color image (left) shows stars in white and radio emission in orange. The radio jets span a distance of some 88,000 light years. Hubble has also imaged the core (top right) and found a gigantic dusty torus, some 800 light years across. This is the first direct observational evidence that astronomers' ideas are correct about the active centers of these galaxies. Sharp as the image is, it still cannot resolve the immediate surroundings of the black hole in the center of the torus.

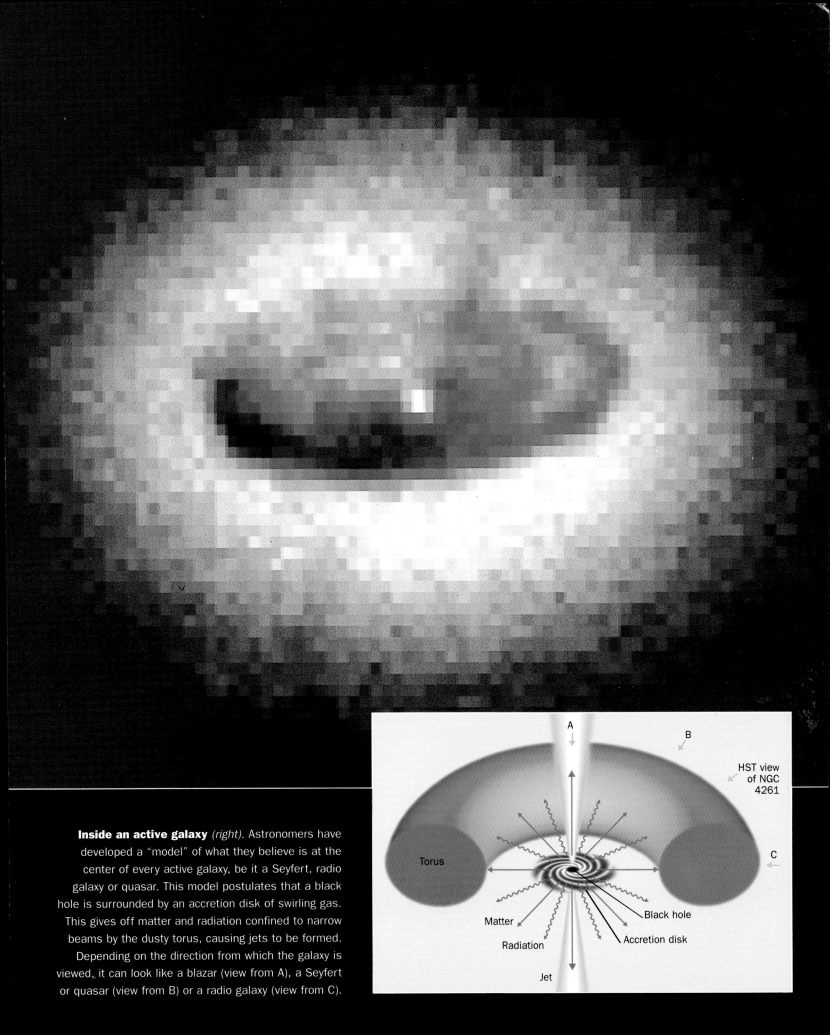

Inside an active galaxy *(right)*. Astronomers have developed a "model" of what they believe is at the center of every active galaxy, be it a Seyfert, radio galaxy or quasar. This model postulates that a black hole is surrounded by an accretion disk of swirling gas. This gives off matter and radiation confined to narrow beams by the dusty torus, causing jets to be formed. Depending on the direction from which the galaxy is viewed, it can look like a blazar (view from A), a Seyfert or quasar (view from B) or a radio galaxy (view from C).

A

B

HST view
of NGC
4261

C

Torus

Black hole

Matter

Accretion disk

Radiation

Jet

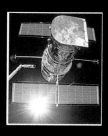

when too much matter crowds in too small a space. Typically, a black hole at the center of an active galaxy will be many millions or even billions of times the mass of the Sun. This will exert a significant influence on the center of the galaxy. The black hole's gravity will be so great that it will rip stars to pieces and devour the gas. As the dismembered star is sucked into the black hole, it will swirl inward, like water going down a drain. This motion will heat the gas to exceptionally high temperatures and cause it to emit powerful radiation. In particular, a jet of radiation-emitting material is squirted out of the center at right-angles to the accretion disk. Surrounding the fury of this central region is a large dusty torus of material, in the same plane as the accretion disk. This blocks our view of the black hole from certain angles and makes the active galaxy look quite different, depending on the vantage point.

Normal galaxy, M51 *(left)* has a dark Y-shape at its center. In the middle of the "Y" is a white spot, as bright as 100 million Suns. This is almost certainly a black hole, but appears white because radiation is being beamed along a jet, directly toward us. The jet raises the possibility that all normal galaxies were once active. Perhaps when the black hole has devoured everything close by, an active galaxy settles down and becomes normal.

In line with this theory, experts think that an active galaxy in which one of its jets points toward us will appear to be a blazar. If the jet is misaligned, the galaxy will appear to be a Seyfert or a quasar depending on the power of the active core. A radio galaxy is an active galaxy in which the equatorial disk blocks our view of the core, and the jets can be seen at right-angles. This theory has not yet been proven beyond doubt because ground-based astronomical technology is, so far, unable to see enough detail in these active galaxies. The Hubble Space Telescope, however, is just able to peer into their centers and give us the first signs that the theory may be correct. Many problems remain to be solved, and one of the principal uses of Hubble's new infrared camera (NICMOS) will be to investigate deep into the centers of active galaxies.

The active galaxy M87 *(left and far left)* in the Virgo cluster. The image of the galaxy (far left) shows a conspicuous jet of material emanating from the core, and an abnormally bright area in the central region. When the Hubble Space Telescope focuses on this bright area (left), it sees a swirling disk of matter. Calculations based on the movement of this disk suggest that a black hole is responsible for the activity in this part of the galaxy.

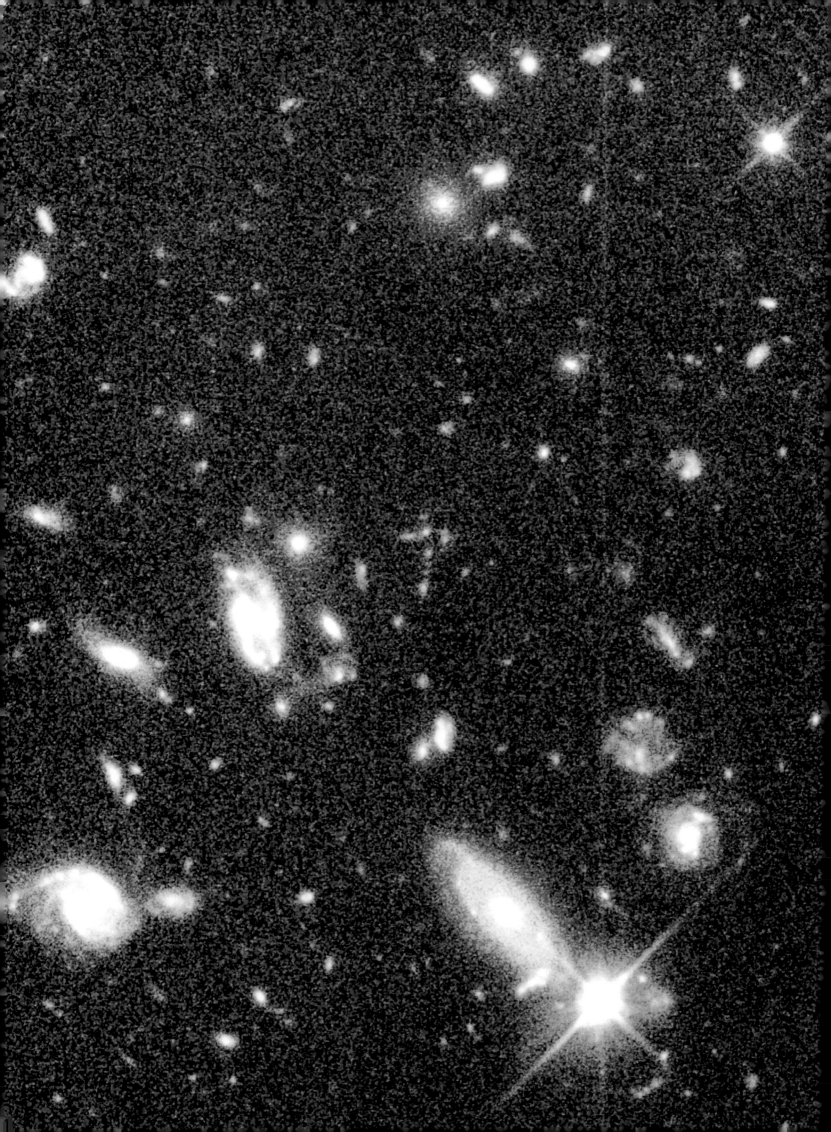

THE
UNIVERSE

*A*LMOST ALL THE HUBBLE IMAGES FEATURED SO FAR *have been of objects quite close to us. Apart from the quasars, everything else has been within a range of a few hundred million light years of Earth. In fact, the universe extends very much farther, at least 15 billion light years in every direction. Most of the universe is a dark wilderness that has yet to be explored telescopically, but the Hubble Space Telescope has allowed astronomers their first glimpse of the celestial bodies that populate the astronomical wilderness.*

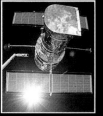

THE EXPANDING UNIVERSE

TRADITIONALLY, STUDYING THE NIGHT SKY is known as astronomy; the science of mapping the heavens and cataloguing the different objects found there. As astronomical surveys become more subtle and sensitive, scientists focus on explaining what the objects in the universe are and how they are formed. This branch of science is known as astrophysics. Cosmologists build on the work of astrophysicists but, instead of trying to understand the individual components of the universe in isolation, they try to explain the evolution of the universe as a whole. Cosmologists ask some of the biggest questions posed by mankind. When was the universe created? How was it created? How does it evolve? How will it end?

In 1929, Edwin Hubble published results to prove his theory that the universe was expanding. This led to the idea that in the distant past the elements in the universe must have been much closer together. This in turn suggested that the universe began its life as an unimaginably compact object, which spontaneously exploded outward. A key prediction in this theory, known as the "big bang", was that the universe should be bathed in a glow of microwave radiation, created at the moment of explosion. In 1964, microwave radiation was indeed detected by scientists Arno Penzias and Robert Wilson. As the universe expands, it carries clusters of galaxies farther and farther away from one another and stretches the wavelengths of light traveling between them. Stretching the wavelength makes the light from a distant galaxy appear more red, and is therefore known as redshift.

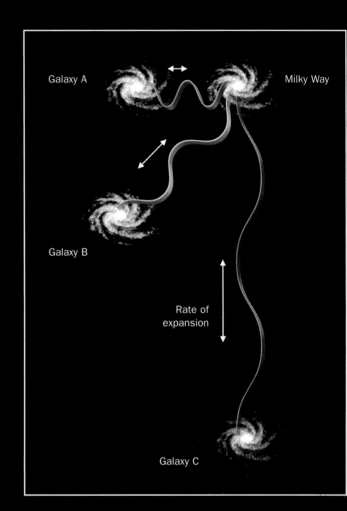

Redshift and the expansion of the universe *(above)*. From our position in the Milky Way, other galaxies appear to move away from us, while we perceive ourselves to be standing still. In fact, it is the universe which is expanding, stretching the space between us and other galaxies. The expansion also increases the wavelength of light as it travels through space, transforming short blue wavelengths into longer red wavelengths. This causes the phenomenon known as redshift; described in its simplest form, the farther away a light source is the more redshifted it is and the redder its light appears. For example, if galaxy B is twice as far away from us as galaxy A, it will have twice as much redshift and its light will appear yellow. If galaxy C is twice as far away again, its light will appear red. Redshift is the visual equivalent of the way the Doppler effect changes the pitch of sound.

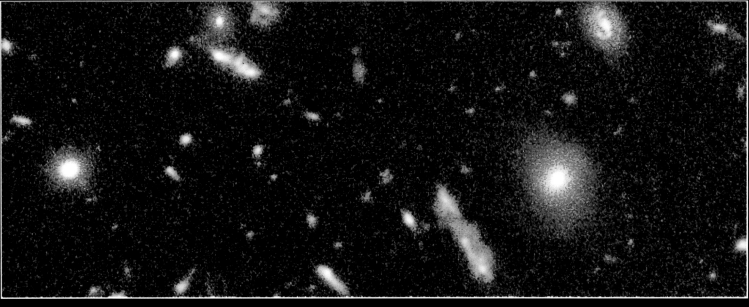

The deep field of the universe *(above)* contains celestial objects, mostly young galaxies, which span billions of years of cosmic time and evolution. The distances from these objects to Earth are so vast that it takes their light millions or even billions of years to reach us. Because their light has been traveling for so long, the objects have changed beyond recognition in the intervening aeons. So we are seeing them as they appeared when the universe was billions of years younger. This gives cosmologists the unique opportunity to see what the universe looked like billions of years ago, by studying objects billions of light years away in space. This field of research has suddenly burgeoned thanks to Hubble's superior ability to see into the very distant universe. Once Hubble has located these incredibly faint objects, experts can do follow-up work using the largest ground-based telescopes.

The expansion of the universe slows, as time goes by, because the clusters exert a gravitational pull on one another, counteracting the pattern of expansion. Cosmologists are trying to find out whether the universe contains enough matter to halt the expansion altogether. If this were the case, shortly after the universe had finished expanding it would begin to collapse under the influence of gravity, eventually ending in a "big crunch", the reversal of the big bang. If the universe does not contain enough matter to halt the expansion, it will simply go on expanding forever.

Cosmological Distance Ladder

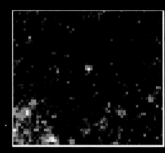

COSMOLOGISTS ARE CONSTANTLY TRYING to determine accurate distances between Earth and the various celestial objects that populate the universe. On the Earth we can measure between two points by traveling from one to the other and logging the distance in between. Since this is not possible when it comes to measuring the distance between celestial objects, we have to devise a number of different methods to measure distances in space.

For the very closest stars, we can use a type of trigonometry known as parallax. This uses the apparent change in a star's position, caused by the Earth moving from one side of its orbit around the Sun to the other, to triangulate the star's position and distance. Parallax does not work well over distances larger than about 100 light years away, because the changes in a star's position become immeasurably small. (Our galaxy is itself about 100,000 light years in diameter, and the distance to the next galaxy is 2.2 million light years). Methods other than parallax must be found if the ladder is to be extended to useful distances.

One of the most accurate methods for determining distance in space is to use a type of star known as a Cepheid variable. These are stars that have left the main sequence and are in the process of becoming red giants. For a brief period, measured in thousands of years, their size oscillates in a regular and predictable manner. The changes in the star's size are accompanied by corresponding changes in its brightness. The greatest advantage

of this instability is that the length of time it takes for the star to pulsate is indicative of the maximum brightness it will reach. If astronomers measure the period of its pulsation, they can calculate a figure for its maximum brightness. We know that the further away the star is, the dimmer it will seem to be. Astronomers can calculate the star's distance from Earth based on the difference between its apparent brightness and its estimated maximum brightness. This method was pioneered by Edwin Hubble and resulted in his calculation of the distance to the Andromeda galaxy. It is now one of the Hubble Space Telescope's primary missions to find individual Cepheid variable stars in ever more remote galaxies in order to calculate their distances from Earth.

Potentially, the most powerful method of determining distances to other galaxies is redshift. It is an easily measurable quantity of any galaxy and is caused by the expansion of the universe. A galaxy twice as far away as another will exhibit twice the redshift. The problem with the redshift scale is that it can only tell cosmologists about the relative distance between galaxies. (It can show that one galaxy is, for example, ten times farther away than

A Cepheid variable star *(below)* in M100 is monitored by the Hubble Space Telescope. This sequence traces the star's pulsation over about 50 days, during which the star doubles in brightness. Its maximum brightness, calculated from its pulsation period, is 2.5 thousand billion times brighter than the star appears to Hubble. This allows its distance from Earth to be calculated as 51 million light years. Since the average spiral galaxy is only 100,000 light years across, the location of the Cepheid variable within M100 is relatively unimportant. When the redshift of M100 is measured, a conversion factor between redshift and distance from Earth can be calculated. The factor is known as Hubble's constant, in honor of Edwin Hubble. The Space Telescope's research should significantly improve the accuracy of Hubble's constant.

 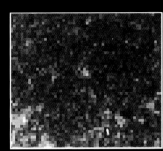  

The spiral galaxy M100 *(right)* in the Virgo cluster is one of the most distant galaxies in the universe in which individual Cepheid variable stars have been detected, thanks to the Hubble Space Telescope. The Cepheid variable stars are all located in the outer reaches of the spiral arms.

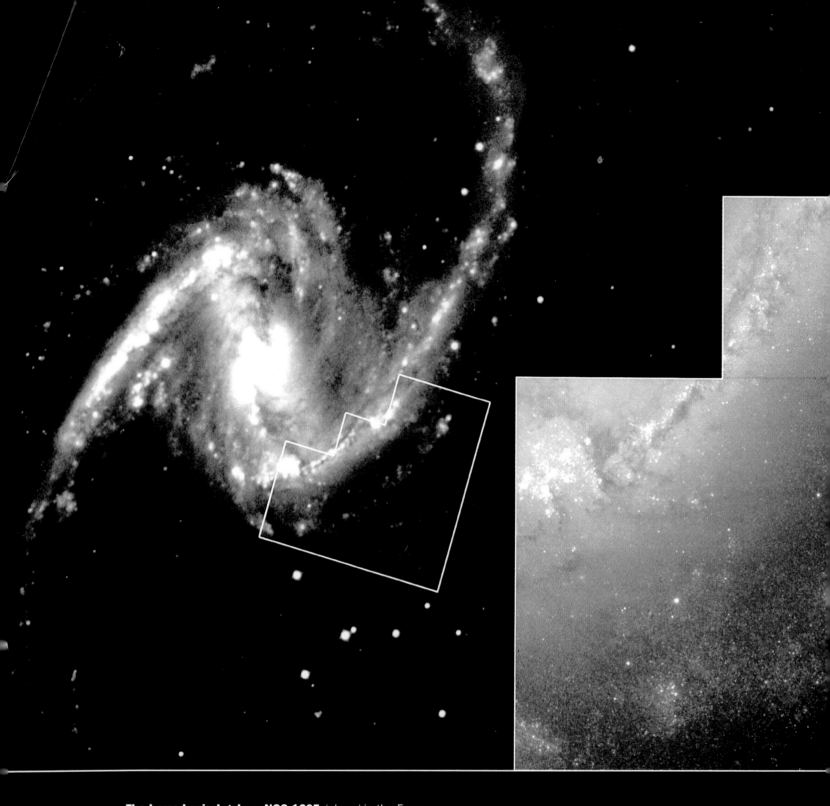

The barred-spiral galaxy NGC 1365 *(above)* in the Fornax cluster of galaxies. The black-and-white image shows the galaxy in its entirety, photographed from a ground-based telescope in the southern hemisphere. The area explored by the Hubble Space Telescope in its search for Cepheid variable stars is marked with a white box. The color image, taken by Hubble, displays the area on a larger scale, and shows a large number of individual stars, as well as other distant galaxies. To date, about 50 Cepheid variable stars have been identified in this image, and careful calculations indicate that

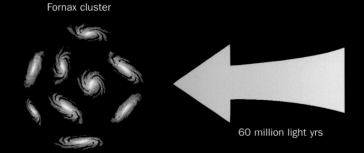

Fornax cluster

60 million light yrs

another). Its biggest drawback, however, is that the absolute distance in light years cannot be determined by using redshift alone. In order to make the redshift scale more useful, we have to find some means of calibrating it by determining the absolute distance to the nearest galaxies to which the redshift analysis is applicable. This is where the Hubble Space Telescope is currently playing a crucial role in refining our initial, crude estimates. By looking for Cepheid variable stars in the Virgo and Fornax clusters, experts can calculate their approximate distance from Earth, and the redshift scale can be calibrated more accurately. Most astromomers believe that redshift will be an accurate indicator of distance, but a few believe that redshift may not be caused solely by the expansion of the universe.

50 million light yrs

Virgo cluster

Milky Way

The estimated distances to the Virgo and Fornax clusters *(left)* have been calculated by Hubble from its position inside the Milky Way, based on analysis of one galaxy within each cluster. Although the precise position of each chosen galaxy in its cluster is not known, Fornax offers a more accurate calibration of the redshift scale since it is a more compact cluster.

A supernova in galaxy M51 *(above)*, viewed in ultraviolet light, provides cosmologists with another way to gauge the distance to M51. Supernovae can be used to estimate distance by measuring the time they take to fade and then applying the same principles used in calculating distances to Cepheid variable stars. Although distance calculations are not as accurate as with Cepheid variable stars, supernovae are so bright that they can be seen across much larger distances in the universe. This supernova was discovered on April 2, 1994. The distance to M51 was subsequently calculated as 20 million light years.

GRAVITATIONAL LENSES

ONE OF THE MOST BIZARRE OCCURRENCES in the universe is a phenomenon known as a gravitational lens. It occurs when a distant quasar or galaxy is positioned almost directly behind a much closer galaxy or cluster of galaxies. As we look from the Earth, the direct light from the distant galaxy or quasar is blocked by the intervening galaxy or cluster, but two or more "mirages" of the distant object appear around the closer object. These cosmic illusions are caused by the gravity of the intervening galaxy or cluster bending the light from the distant object, which we would normally never see, into our line of sight. Our brains are so used to light traveling in straight lines that we interpret what we see as meaning that there are several distant galaxies or quasars on either side of the nearby galaxy or cluster, rather than just one directly behind it. Sometimes the galaxy or cluster causing the lensing is too dim to be seen. Only with careful scientific analysis has it been possible to tell that, what appears to be several distant objects are, in fact, all mirages of the same one.

Hubble's 100,000th shot *(top)* shows a potential gravitational lens. The object in the center is a quasar, with a star to its right. Above the quasar is a foreground galaxy. If this galaxy were directly between us and the quasar, it would act as a gravitational lens. This is what has happened *(above)*. The objects top and bottom are mirages of the same quasar, with the lens in the center.

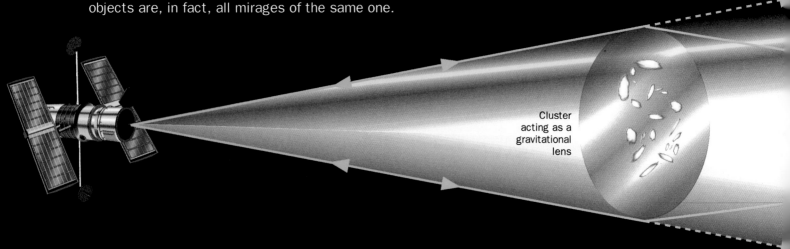

Cluster acting as a gravitational lens

Gravitational lensing *(above and left)*. Galaxy cluster 0024+1654 (above) acts as a gravitational lens to a young galaxy behind it. The young galaxy appears as blue arcs in a ring around the cluster's center. The length of the arcs depends on how precisely aligned the cluster is with our line of sight to the young galaxy. If the alignment were perfect, the galaxy would appear as an unbroken ring of light, an "Einstein ring". The diagram shows how Hubble views light through a gravitational lens. Light from the distant galaxy is bent by the intervening cluster, acting as a focusing lens. This creates a visual effect, making it seem as if there is a ring of objects (in reality a mirage) beyond the gravitational lens.

Mirage

Young
distant galaxy

EVOLUTION OF GALAXIES

SOON AFTER THE BIRTH OF THE UNIVERSE, matter was spread fairly uniformly throughout the cosmos, not grouped together to form objects such as stars and galaxies. Cosmologists are trying to work out how matter gradually aggregated, first into clumps and then into galaxies. The first clues about how galaxies evolved came in the early 1970s when cosmologists caught their first sight of faint blue galaxies in the very distant universe. The Hubble Space Telescope has given us a good look at these mysterious, faraway galaxies. Because their light has taken so long to reach us, we are seeing these galaxies as they appeared millions of light years ago, a phenomenon known as "look-back time".

Galaxies in the universe today fit into Hubble's classification scheme, but these older galaxies do not. They are irregular in shape and burn brilliantly because of unusually active star formation.

More importantly (from the point of view of cosmic evolution) they vastly outnumber the normal galaxies in the universe today. Some experts believe that we cannot see this type of galaxy around us today because they have exhausted the supply of star-forming gas and, consequently, have faded below limits of detectability.

In an attempt to trace the evolution of normal galaxies, the Space Telescope has identified three clusters of galaxies at cosmologically significant distances. The most distant cluster (so far) is approximately 12 billion light years away in the constellation Sculptor. The cluster contains 14 galaxies and, because of the effects of look-back time, shows us how these galaxies appeared approximately 12 billion years ago. They look nothing like the recognizable spirals of today, although some resemble ellipticals. The second cluster under Hubble's scrutiny is closer, about nine billion light years away, in the constellation Serpens. By this stage in their evolution, many of the galaxies begin to look as if they might become spirals.

Irregular blue galaxies *(left)* are easily visible in this image. The galaxies themselves, though small, outnumber normal spirals and ellipticals many times over. Evidence suggests that they were the dominant form of galaxy several billion years ago but have faded to obscurity in today's universe.

These faint clumps of stars *(right)* are some of the objects found by Hubble during its search for galaxy building blocks. Each one is some 11 billion light years away. If our galaxy were shown at the same distance and scale, it would stretch across three of these images. This implies that clumps must have merged together in order to make today's galaxies.

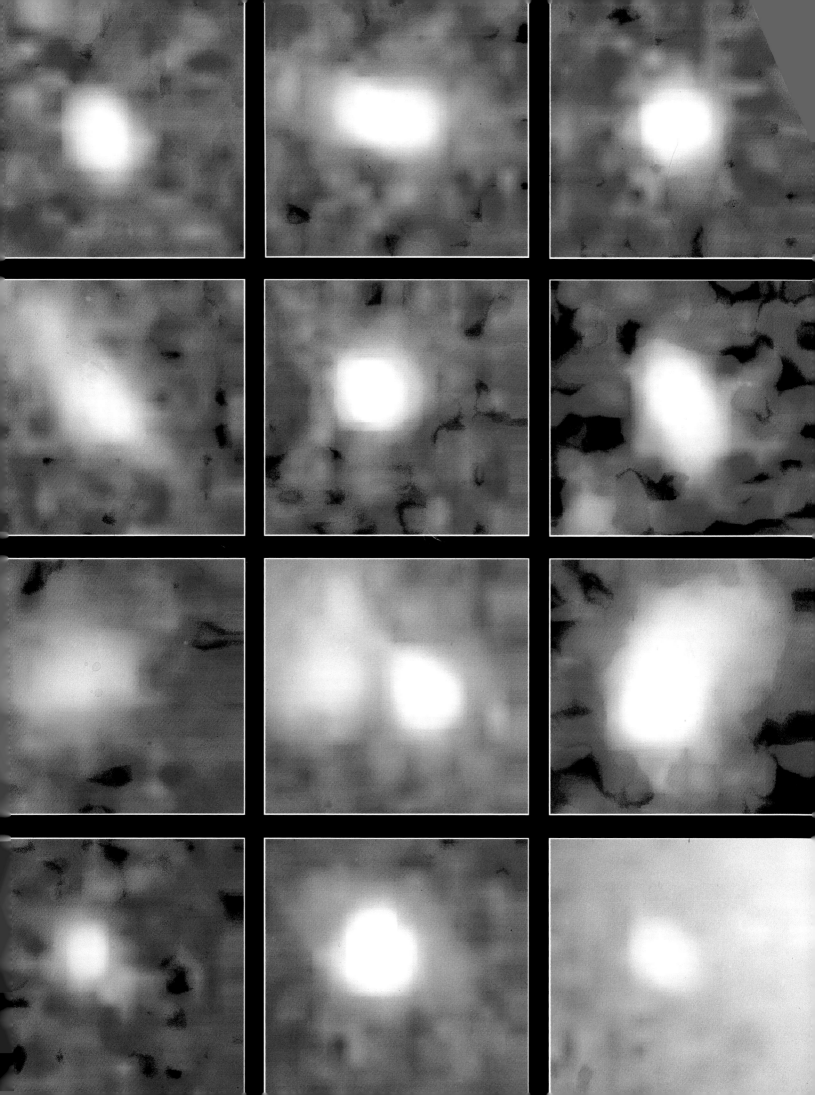

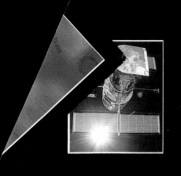

Obvious candidates for ellipticals are also present. Many others, however, appear to be fragmentary objects. Perhaps they are the building blocks of today's galaxies. In follow-up observations to investigate the concept of galaxy building blocks, cosmologists found 18 young galaxies in a region of space no wider than two million light years (the distance between the Milky Way and the Andromeda galaxy). Each building block is between 2,000 and 3,000 light years in diameter, which makes them too small to be galaxies. This discovery supports the theory that they are tiny building blocks of matter, which will merge to form larger galaxies in the following few billion years. It also fits with the theory that, shortly after the big bang, the universe's initially smooth distribution of matter fragmented in parts and formed tiny galaxies. These grew into much larger galaxies, either by merging with others or by accumulating matter from their surroundings. The final cluster examined in Hubble's survey of galaxy

Some 18 very young galaxies *(right)* are visible in Hubble's image of a region near the constellations Hercules and Draco. The red and green exposures were photographed in June 1994 and stored on computer to await the blue exposure, captured in June 1995.

These intriguing galaxies *(below)* were captured by Hubble on a mission to find faint blue galaxies. The one on the left is a prime example of a nascent spiral, while the one on the right is still an amorphous clump.

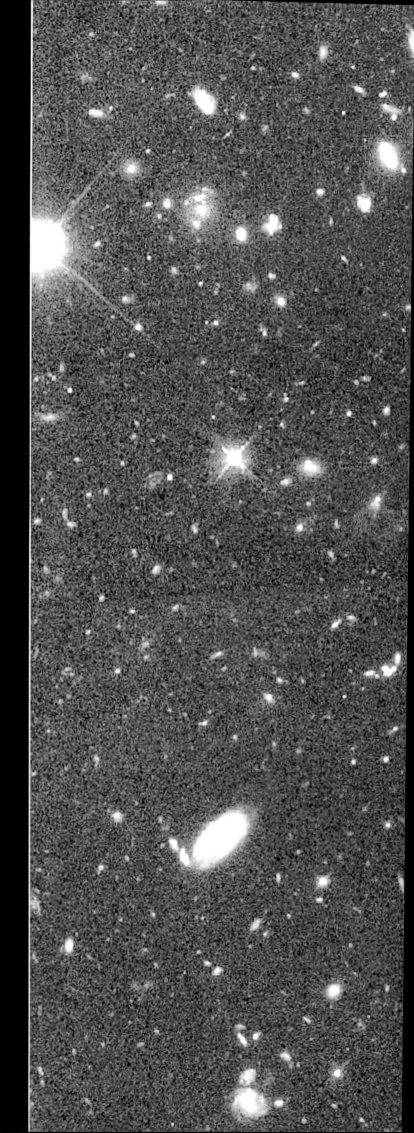

| Today
14 billion yrs old | 5 billion light yrs away
9 billion yrs old | 9 billion light yrs away
5 billion yrs old | 12 billion light yrs aw[a]
2 billion yrs old |

Elliptical galaxy

Spiral galaxy

The typical evolution pattern for normal galaxies *(above)* can be seen in these two sequences of images. All these galaxies are in distant clusters, and appear as they would have done up to 12 billion years ago, depending upon how far away they are. Their age, calculated from the look-back time, appears above each pair of images. Based on Hubble Space Telescope estimates that the universe is 14 billion years old, the most distant cluster lets us see spiral and elliptical galaxies as they were two billion years after the big bang. The evidence suggests that elliptical galaxies change very little during their evolution. In contrast, spiral galaxies appear to change greatly during the course of their development.

evolution is one that had already been catalogued. CL0939+4713 is five billion light years away and contains galaxies that are more closely reminiscent of today's spirals and ellipticals. An intriguing feature of this cluster is the greater proportion of spiral galaxies when compared with the usual proportion in clusters today. This bolsters the belief that, as time goes by, spiral galaxies merge or collide with one another, producing elliptical galaxies. In the future, spiral galaxies will continue to merge, so the number of ellipticals in the universe will increase. Not even our own galaxy will be spared. The Milky Way is currently on course for collision with the Andromeda galaxy in the distant future!

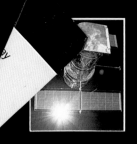

THE HUBBLE DEEP FIELD

THE SPACE TELESCOPE'S crowning achievement so far is the Hubble Deep Field. This historic image was captured over Christmas 1995 when, for ten consecutive days, Hubble peered at the same region of the universe, collecting as much data as possible. Between December 18 and December 28, it took 342 images of the same field, a tiny patch of sky above the "Big Dipper". This area was chosen because a minimum number of nearby stars and galaxies obscured Hubble's view. The images were combined, using computer technology, to produce the Hubble Deep Field. The result can be regarded as a "keyhole" view through the entire universe. Estimates suggest that there are over 3,000 galaxies on this image, most of them so dim that they are four billion times too faint to be seen by the naked eye! The Deep Field is being studied by research teams around the world. It shows them the whole cosmological evolution of the universe in a single amazing image. All they have to do now, is to decipher what it tells us.

A portion of the Hubble Deep Field *(right)* showing billions of years of cosmic evolution in a revolutionary picture of the universe. The diagram (inset bottom) shows Hubble's line of view. Although this part of the Deep Field only spans a diameter of about 1/30th that of the full Moon, it provides astronomers with a wealth of data about cosmic evolution and the development of galaxies. The objects are representative of what the universe contained at each and every stage of its development. Only the bright point of light near the center of the main image is a star, all the rest are galaxies, thousands of them never seen before. One of the most distant galaxies identified so far is highlighted (inset top).

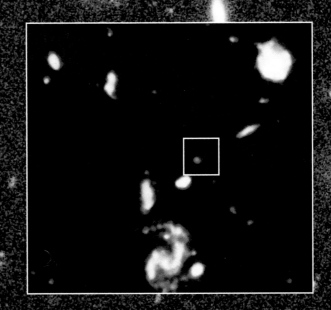

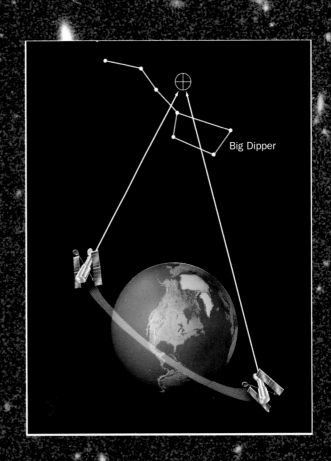

Big Dipper

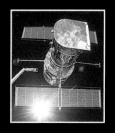

THE

FACTFILE

*T*HE FACTFILE CONTAINS HELPFUL REFERENCE INFORMATION
to supplement the main content of the book.
In particular, the Keywords section provides
definitions of technical terms used
in the body of the text.

1889 Edwin Hubble is born.

1923 Hermann Oberth writes about a telescope in Earth orbit.

1929 Edwin Hubble publishes evidence that the universe is expanding.

1946 Lyman Spitzer writes about the possibility of a telescope in space, and anticipates many of
the Hubble Space Telescope's fields of study.

1953 Edwin Hubble dies.

1957 Sputnik 1 is launched and the space age begins in earnest.

1966 Lyman Spitzer chairs the first meeting of the "Ad Hoc Committee on the Large Space Telescope
(LST)" to discuss putting an 118-inch (3 meter) telescope into orbit.

1969 The "Ad Hoc Committee" publishes a report entitled "Scientific uses of the Large Space
Telescope". Apollo 11 lands on the Moon.

1972 The concept of the Space Telescope Institute is proposed.
The Space Shuttle project begins.

1974 The "Science Working Group for the LST" decides to downsize the telescope's primary mirror
in order to achieve a realistic budget.

1977 NASA begins the Space Telescope project after the US Government approves its budget.
Launch date is set for 1983. NASA launches Voyagers 1 and 2.

1979 Actual construction of the telescope begins. Voyagers 1 and 2 make fly-bys of Jupiter.

1980 Voyager 1 makes a fly-by of Saturn.

1981 The maiden flight of the Space Shuttle Columbia. Voyager 2 makes a fly-by of Saturn.

1983 The Space Telescope is named the Hubble Space Telescope (HST) after Edwin Hubble.
Launch is delayed until 1986.

1986 Voyager 2 makes a fly-by of Uranus. The tragic explosion of the Space Shuttle Challenger
further postpones the launch of the HST.

1989 Voyager 2 makes a fly-by of Neptune. NASA launches the Galileo probe to Jupiter.

1990 HST is launched from the Space Shuttle Discovery. The primary mirror is discovered to be faulty.

1993 HST's first servicing mission. The Space Shuttle Endeavour docks with the telescope, and its
astronauts install and replace key instruments to correct the faulty mirror.

1995 Galileo probe arrives at Jupiter.

1997 HST's second servicing mission. The Near-Infrared Camera and Multi-Object Spectrometer
(NICMOS) and the Space Telescope Imaging Spectrograph (STIS) were installed.

1999 HST's third planned servicing misson. The Advanced Camera will be installed, and other
servicing work will be carried out.

2002 HST's fourth planned servicing mission. Instruments will be replaced and updated.

GROUND-BASED TELESCOPES

Even though the Hubble Space Telescope is helping astronomers to make remarkable progress, it has not, by any means, made ground-based telescopes redundant. Ground-based astronomy is still as important as it ever was and, indeed, many new telescopes are currently on the drawing board and under construction. Hubble will never detect radio waves, microwaves or high-energy emissions such as X-rays and gamma rays, as the Space Telescope was not designed for observation at those wavelengths. This research will remain the province of ground-based radio telescopes and high-energy detection satellites.

Radio telescopes are the premier tool of astronomy after optical instruments. They are enormous dish systems, often constructed side-by-side and used, in tandem, to produce incredibly precise images of the radio emissions from celestial sources. When a pair (or multiple pairings) of radio telescopes are used together, they simulate the capability of a radio telescope with a much larger dish. Using dishes in tandem is known as interferometry. It has been pioneered in radio astronomy and is now used at many of the most important radio observatories around the world. The larger the number of dishes used in the interferometer, and the greater the distance between the dishes, the higher the quality of the images obtained.

Radio telescopes *(below left)* bring radio waves to a focus by reflecting them from a large curved dish upward toward a receiving antenna. This antenna is held above the dish by support struts. Electrical signals produced in the antenna are transported by wires to the electronic units for analysis. The Very Large Array near Socorro, New Mexico, *(below)* uses a total of 27 radio telescopes, which can be moved along rails to allow different configurations and pairings. Using a number of telescopes in tandem, the Very Large Array works as an interferometer, producing incredibly detailed pictures.

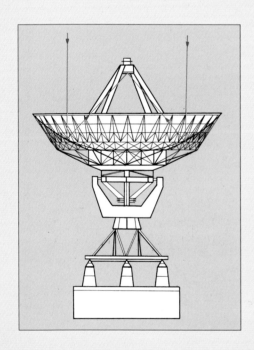

For a long time, it seemed that a 200-inch (5-meter) primary mirror was just about the biggest that could be constructed to an acceptable standard for optical telescopes. Advances in optical technology have since allowed astronomers to employ even larger single mirrors and, at present, a new generation of 315-inch (8-meter) class telescopes is being constructed. However, the largest telescopes in the world today are a pair of identical twins known as Keck I and Keck II. These were constructed on Mauna Kea, an extinct volcano on Hawaii, 2.5 miles (4 kilometers) above sea level.

Optical telescopes (below) exist in a wide range of types. The largest optical telescopes in the world, however, all tend to work in a similar way. They use curved mirrors to collect and focus light. The primary mirror is the bigger mirror and focuses incoming light back up the telescope toward a secondary mirror, which reflects the light back down the telescope and completes the focusing process. The four domes of the revolutionary Very Large Telescope (below right) are shown while still under construction in December 1996.

Each Keck possesses a primary mirror which is a mosaic of 36 hexagonal mirrors, all held in place by computer controlled grippers. This ingenious design enabled the technicians to side-step the problems inherent in large single mirrors to produce a mosaic primary mirror of no less than 394 inches (10 meters) in diameter.

Optical telescope makers are also borrowing the interferometry techniques of the radio astronomers. Some of the 315-inch (8-meter) class optical telescopes will be linked up so that they will be capable of working together in the same way as radio telescopes. It is envisaged that the two Keck telescopes will be able to work like this. The most ambitious plans using 315-inch (8-meter) class multiple optical telescopes rest with the European Southern Observatory's four-telescope cluster, the Very Large Telescope (VLT), under construction at the Paranal Observatory in South America.

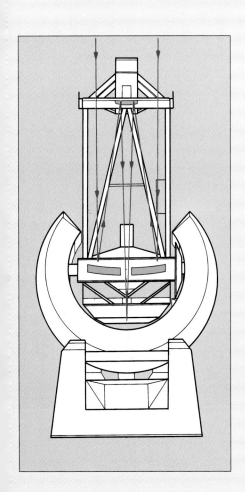

KEYWORDS

absorption nebula
A dark region of dust and gas, denser than the **interstellar medium**, that contains no stars but absorbs the light from objects behind it. Some absorption nebulae are large sprawling dusty lanes that can be seen linked with the bright spiral arms in spiral and barred-spiral galaxies. The three pillars in the Eagle nebula, M13, are absorption nebulae silhouetted on the bright emission portion of the Eagle nebula. See **nebula**.

accretion disk
A disk of dust and gas around newly-forming stars and **black holes**. When forming young stars, parts of the **absorption nebula** collapse while it remains in orbit around the center of the galaxy; the material falls inward, forming an accretion disk with a **protostar** in its center. The accretion disk around a black hole is more active because the force of gravity there is much greater. Anything in orbit around the black hole is pulled towards it and ripped to pieces.

active galactic nucleus
The tiny central region of a **galaxy** that emits extremely high levels of radiation. One theory suggests that the heart of an active galactic nucleus is a **black hole** with an **accretion disk** the diameter of our **Solar System** around it. This nucleus is one of the most powerful objects in the universe. As material is ripped to shreds and funneled around the black hole, it is heated and emits copious quantities of radiation. This radiation is what causes observers to term the galaxy active.

active galaxy
Any galaxy that has an **active galactic nucleus**. Active galaxies are divided into four types: **Seyfert galaxies**, **radio galaxies**, **quasars** and **blazars**. One theory suggests that all types of active galaxies are the same, viewed from different angles.

aperture door
A cover fitted to the top of the Hubble Space Telescope to keep out light and unwanted objects when the telescope is not taking an exposure.

asteroid
Any of the small rocky minor bodies that orbit the Sun. Asteroids are primarily found in a zone between Mars and Jupiter, but some have eccentric orbits that may cross the orbit of the Earth. The largest known asteroid is Ceres, with a diameter of 584 miles (940 km).

astronomy
The study of the universe in which we live. Astronomers of antiquity charted celestial objects and described their appearance. Today charting celestial objects is known as astrometry, and trying to understand their nature is known as **astrophysics**. See **cosmology**.

astrophysics
The study of the inner workings of celestial objects according to the known laws of physics.

atmosphere
The gaseous envelope around a planet. The Earth's atmosphere is relatively thin and is composed mostly of nitrogen and oxygen; that of Venus is composed of carbon dioxide and sulfuric acid clouds, while Mercury has no real atmosphere at all. Jupiter, Saturn, Uranus and Neptune have very dense atmospheres.

aurora
A glow of radiation in the sky, close to a planet's magnetic pole, caused by electrically charged particles usually expelled from the Sun and caught in the magnetic fields surrounding planets. Aurorae are not just caused by particles from the Sun. A different kind of aurora is produced by volcanic eruptions on Jupiter's moon, Io, which cause particles to be expelled; they are caught in Jupiter's magnetic field and funneled into its upper atmosphere, creating aurorae.
See image on page 115.

barred-spiral galaxy
A type of **galaxy** in which a nucleus of old stars is joined by a bar-like structure of dust, gas and stars to the spiral arms containing young stars (see page 78).
See image on page 121.

big bang
A theory that states that the universe came into being in an instantaneous event that happened between 15 and 20 billion years ago. Matter was created at once, and gradually evolved into stars and galaxies, while the universe itself has continued to expand. The laws of physics describe everything that has occurred after the first fraction of a second.

big crunch
A theory that the universe will end by collapsing in upon itself. The expansion of the universe is being slowed by the force of **gravity** between clusters of galaxies. If the universe contains sufficient matter, its gravitational fields will be strong enough to halt expansion, and contraction will begin. As space contracts, its temperature will increase. This could lead to another **big bang**. Some experts believe that the universe will begin expanding again; others think it will cease to exist.

billion
1,000,000,000 – one thousand million.

binding energy
The energy given out when two atomic nuclei are joined together by the process of nuclear **fusion**. The release of binding energy as hydrogen atoms fuse in the core of a **main sequence star** is what causes the star to shine.

black hole

An object so dense that even light cannot escape its gravitational field. Black holes cannot be directly observed, but they can be detected by the presence of X-rays emitted from **accretion disks** that are formed when stars fall into black holes. There are two main types of black hole. One is formed when a massive star explodes as a **supernova**. If its collapsing core is more than three times the mass of the Sun, it will become a black hole. The constellation Cygnus is thought to have a black hole of this type. The second type of black hole is a supermassive one at the heart of an **active galaxy**, millions or even billions of times the mass of the Sun. It forms as the central concentration of matter in a galaxy.

blazar

A type of **active galaxy** related to a **quasar**. Blazars appear to vary in brightness. They are thought to be very distant and bright **active galactic nuclei**.

blue supergiant

Any star in the upper left corner of the **Hertzsprung-Russell diagram** (see pages 52-53). Blue supergiants have high mass, high surface temperature (over 20,000 K) and a short lifespan (up to 10 million years). Spiral arms and young galaxies appear blue because they contain large numbers of blue supergiants.

brown dwarf

A large gaseous object, neither a **star** nor a **planet** (similar in composition to Jupiter but much more massive), which has yet to reach the critical mass required to ignite nuclear **fusion** in its core and turn it into a star. A collapsing gas cloud produces only a few high-mass stars but many low-mass stars, so there may be a huge number of brown dwarfs undetected in the universe. They are difficult to locate because they are small and dim.

Cepheid variable star

A **star** (typically yellow in color) that has left the **main sequence** represented on the **Hertzsprung-Russell diagram** (see pages 52-53) and is well on its way to becoming a **red giant**. Unstable nuclear **fusion** in the star's core makes it pulsate, producing variations in the star's dimensions, temperature and luminosity. All of these properties vary within a regular, repeating period of between 3 and 50 days. The length of this period is related to average brightness of the star by a mathematical equation known as the period-luminosity relationship. Surface temperatures can vary by as much as 1,500° C, while the radius varies by about 10 to 30 percent, and luminosity changes by a factor of 2.5. Cepheids are named after the prototype star, Delta Cephei. They occupy a region on the Hertzsprung-Russell diagram known as the instability strip. The period-luminosity relationship makes these stars useful for gauging distance to galaxies.

charge-coupled device (CCD)

An electronic device for taking pictures of celestial objects. Charge-coupled devices are sensitive to almost every ray of light striking their surface and so take images, especially of faint objects, easily and quickly. CCDs can also convert light into electrical signals. This makes their output readily available for computers to analyze the images.

colliding galaxy

Any **galaxy** that is in the process of colliding with another. The force of the collision usually distorts the shape of one or both participants and results in an intense bout of star formation. This is why the **starburst** phenomenon is often associated with colliding galaxies. The starburst is triggered by the collision of **giant molecular clouds** within the galaxies, causing shock waves and compression of the **interstellar medium**.

color

A manifestation of the wavelength of **visible light**. The human eye is sensitive to electromagnetic radiation with wavelengths between 350 and 700 nanometers (one billionth of a meter). Within this range, the light appears as different colors. Blue light has the shortest wavelengths in the visible spectrum, less than 400 nanometers, while red light has the longest, over 600 nanometers. Yellow light, such as that given out by the Sun, has a wavelength of 550 nanometers.

comet

Any small icy object in the **Solar System**. Most comets are thought to lie in distant regions known as the Kuiper belt and the Oort cloud. Comets so far away from the Sun are difficult to observe, because they are so small. Gravitational instabilities in their **orbits** (or the pull of large planets such as Jupiter) can draw them toward the Sun, where they can be studied. As a comet approaches the inner Solar System, it receives increased amounts of solar radiation, melting the ice on its surface. This is what creates the comet's "tail". Some comets from the Kuiper belt, such as Halley's comet, have highly elliptical orbits that bring them into the inner Solar System once every 100 years or so. With each passage through the Solar System, they lose more mass, until eventually they break up altogether. In other cases, a comet's orbit may be so changed that it escapes the Sun's gravity and wanders into interstellar space.

constellation

A pattern of stars, or the area of the night sky in which a particular pattern occurs. As seen from Earth, the sky contains 88 constellations of varying size; Hydra is the largest, Crux the smallest. Objects within the constellation boundary, no matter how far away in space, are considered to belong to the constellation, such as the Virgo cluster of galaxies.

corrective optics

A system of lenses designed to correct the distorting effects of the **spherical aberration** in the primary mirror of the Hubble Space Telescope. See **COSTAR**.

cosmology

The study of the universe as a whole. Cosmology is principally concerned with how the universe became the way it is, the way it has changed over the course of its evolution, and what will happen to it in the future.

COSTAR

The optical system fitted into the Space Telescope's instrument cluster to correct the **spherical aberration** of the **primary mirror** before light enters any of the surrounding instruments. COSTAR

(Corrective Optics Space Telescope Axial Replacement) consists of five mirrors, each of which is approximately the size of a coin. They have been ground to refocus the aberrated light perfectly. Not all the mirrors are used during each observation; the instrument making the observation determines which set of mirrors the light is directed toward. *See images on page 115.*

Doppler effect

An everyday physical phenomenon that causes the wavelength of sound to be altered, because either the source of the sound and/or the listener is moving. When the source and listener are approaching each other, the wavelength of the soundwaves is "squashed" closer together, which means that the pitch increases. When the source and the listener are moving apart, the wavelength is "stretched", so that the sound drops in pitch. **Redshift** is the visual equivalent of the Doppler effect, and works on the same principle.

eccentricity

The measure of an orbit's deviation from a circular shape. Eccentricity is measured as a number from 0 to 1. A circular orbit corresponds to an eccentricity of 0. Increasing the elliptical shape of the orbit increases the eccentricity. An eccentricity of 1 corresponds to a parabolic orbit.

ecliptic

The plane of the **Solar System**. All the planets in the Solar System orbit in the same flat plane; so do most of the asteroids. This is a result of our Solar System forming in a flat **accretion disk** around the protosun 4.5 billion years ago.

Einstein ring

A very rare visual phenomenon, produced by a **gravitational lens**. Under normal circumstances a gravitational lens will produce two or more "mirages" of a distant object positioned behind it. Less frequently, the alignment between the lens and the distant object may be close enough to produce semi-circular arcs instead of individual mirages (see pages 98–99). However, when the lens, the observer and the distant object are in perfect alignment, the gravitational lens produces a mirage that forms a complete circle, an Einstein ring, around the lens.

Einstein's theories of relativity

Two inter-related theories published by the Nobel Prize-winning physicist Albert Einstein at the beginning of the twentieth century. "Relative" means that physical properties such as mass and velocity are not absolute, but can vary according to circumstances. This was a startling, revolutionary idea in its time. The special theory of relativity, published in 1905, sought to explain how the universe would appear to different observers who were in different states of motion from one another. A key principle was that velocity could only be measured relative to another object, so that no experiment in an isolated laboratory could determine the speed of the Earth's movement through space. Only by observing an object not on the Earth – such as the Sun – can we determine our velocity relative to the object; that, in turn, is almost certainly different from our velocity relative to any other object. In 1915, Einstein extended the theory to include accelerating objects. This new work was called general relativity and had far-reaching consequences because it provided a good mathematical tool with which to study **gravity**. The theory of general relativity is one of the foundations of **cosmology**.

elliptical galaxy

A type of **galaxy** in which the constituent stars are all contained within a central grouping. There are no spiral arms, and the majority of stars are older and more evolved than in a **spiral galaxy**. An elliptical galaxy can range in shape from a perfect sphere to the shape of an American football, depending on how the galaxy was formed. One theory is that elliptical galaxies are formed when spiral galaxies collide or merge.

emission nebula

Any cloud of interstellar gas that is giving out electromagnetic radiation. In **star-forming regions**, ultraviolet light from **blue supergiant** stars causes the surrounding gas to emit light. In **Herbig-Haro objects**, the emission is produced by shock waves traveling through the **interstellar medium**. Emission nebulae are also found around **supernovae** and low-mass **stellar remnants** such as **white dwarfs**. In general, emission nebulae are usually isolated regions of much larger interstellar clouds. See **nebula**. *See image on page 115.*

equator

The theoretical plane that bisects a rotating celestial object into two hemispheres and is at right angles to the rotation axis.

Faint-Object Camera (FOC)

A highly sensitive instrument in the Hubble Space Telescope's instrument cluster that photographs very faint celestial objects. The camera is sensitive both to **visible light** and to some **ultraviolet radiation**. After light from the telescope passes into the camera, it is guided through the FOC by two other mirrors and through a set of colored filters. The light is then passed to an image intensifier and finally "photographed" using a variation of a television camera.

Faint-Object Spectrograph (FOS)

A spectrograph designed to observe celestial objects at wavelengths between **ultraviolet** and **near-infrared**. It was originally in Hubble's instrument cluster, but was replaced by the **Near-Infrared Camera and Multi-Object Spectrometer**. The FOS can split the light from faint or distant objects into a spectrum so that their chemical composition and chemical state can be studied. Light is split into a spectrum by a grating, which operates like an ordinary prism but is easier and more accurate to use. The spectrum is recorded for transmission to Earth.

false-color image

An image that has been artificially colored in ways that do not match how the eye would actually see the object when it was photographed. Completely "wrong" colors may be used in order to highlight subtleties of detail that are otherwise too difficult to see. Sometimes false colors have to be used to enhance wavelengths of radiation that are normally outside the visible region of the spectrum. See true-color image. *See image on page 115.*

first light
The first time that a new telescope is used to observe the night sky.

fusion
A nuclear reaction in which two atomic nuclei join together to form a new one. In the center of a star, for example, hydrogen nuclei are exposed to extremely high temperature and pressure, which cause them to fuse easily and rapidly, eventually producing helium in a four-stage process. At each stage, the binding energy of the nuclei involved is given out as radiation.

galaxy
A collection of stars, dust and gas, held together by the force of its constituents' **gravity**. Galaxies exist in a wide variety of shapes and sizes. One of the most popular classification systems is shown on the Hubble tuning fork diagram (see page 78). Galaxies are described as **elliptical galaxies**, **spiral galaxies**, **barred-spiral galaxies** or **irregular galaxies**. The smallest can contain fewer stars than a typical **globular cluster** (perhaps a few hundred thousand) and are known as dwarf irregular or dwarf elliptical galaxies. The largest galaxies, containing trillions of stars, are vast elliptical galaxies, which are collections of older stars often found at the center of a cluster (a group of galaxies held together by gravitational forces). Newer stars and **star-forming regions** tend to be found in spiral galaxies. There are also galaxies in which an energetic process is taking place at the center. These are known as **active galaxies**.

Galileo probe
A NASA space probe that was sent to Jupiter in 1990 to conduct an extensive study of the giant planet and its moons. The Galileo probe was not designed to make a brief fly-by of the planet such as its predecessors, the **Voyager probes**, did. Instead, it was designed to be captured by the gravitational field of Jupiter and orbit the planet. Its planned flight-path was chosen so that the probe would periodically make close passes of Jupiter's moons, gathering information about them. Galileo was the first space probe to make an extensive study of a part of the outer **Solar System**.

geostationary orbit
A **satellite**'s fixed **orbit** at an altitude of approximately 22,300 miles (35,900 km) above the Earth's surface. At this height, the orbital period is exactly the same as the rotational period of the Earth, so the satellite completes one orbit as the Earth rotates once about its axis. In this way, the satellite always remains above the same geographical location, which allows it to have a constant, unbroken stream of communication with its ground station. Typically, geostationary orbit is used for communication satellites.

giant molecular cloud
A vast area of interstellar gas that stretches through space over a distance of several hundred light years. The name derives from the fact that a large proportion of the constituent gas atoms will have joined together to form molecules that can be detected with **radio telescopes**. A giant molecular cloud can contain up to ten million times the mass of the **Sun** in gaseous material. Almost certainly, isolated regions of the cloud will have already experienced gravitational collapse, formed **stars** and be shining as **emission nebulae**.

globular cluster
A spherical collection of old stars, one of several such clusters in orbit around an **elliptical galaxy** or, less commonly, a **spiral galaxy**. It is thought that globular clusters form at the same time as the galaxy that they orbit. Star formation has stopped in most globular clusters; in fact, it is assumed that it occurs while the globular cluster is forming and ceases soon afterward. This means that the stars in a globular cluster are approximately the same age. They play a vital role in helping us to understand the evolution of stars. *See image on page 115.*

Goddard High-Resolution Spectrograph (GHRS)
A spectrograph designed to work in the ultraviolet region of the spectrum. It was originally in Hubble's instrument cluster, but was replaced by the **Space Telescope Imaging Spectrograph**. The GHRS covered the same ultraviolet range as the **Faint-Object Spectrograph** but was

capable of more precise spectra. The FOS could be a survey instrument, with the GHRS following up in more detail.

gravitational lens
Any massive celestial object, such as a galaxy or a cluster of galaxies, which bends the light coming from a more distant celestial object. This distortion of the light usually results in more than one image of the distant object being visible to the observer (see pages 98–99). If the alignment of the observer, gravitational lens and distant object is perfect, a ring of light, known as an **Einstein ring**, will be visible around the gravitational lens.

gravity
A fundamental force of nature, gravity acts between two or more objects with mass, attracting them to each other. Gravity shapes the galaxies and clusters of galaxies in the universe. In fact, it is the dominant force over cosmological distances. The best description of how gravity works within the universe is set out in **Einstein's theories of relativity**; specifically the general theory of relativity.

Great Dark Spot
A large circular storm in the atmosphere of Neptune, the Great Dark Spot was discovered by the Voyager 2 space probe during its 1989 fly-by of the planet. See **Voyager probes**.

Great Red Spot
An enormous circular storm that exists in the atmosphere of the planet Jupiter at a latitude of 22° south. It has been observed on Jupiter since the invention of the telescope and appears to be a permanent or semipermanent feature of the atmosphere.

Greenwich Mean Time (GMT)
The time, measured from the prime meridian at Greenwich Observatory, London, UK. GMT (also called universal time) is the timescale by which all astronomical measurements are referenced.

ground-based telescope
Any telescope situated on Earth. Ground-based telescopes contend with several difficulties that do not affect the Hubble

Space Telescope. First, they can only observe at night, whereas Hubble can observe for 24 hours each day. Secondly, they can only observe when there is clear weather. Hubble is above the atmosphere and cannot be obscured by cloud. Finally, even in good conditions, ground-based telescopes cannot perform to their theoretical maximum, because incoming starlight is distorted by the atmosphere. However, the great advantage of ground-based telescopes is that they are far more accessible if anything goes wrong.

Herbig-Haro object

An **emission nebula**, produced when a jet of high-speed gas shoots from a **protostar** colliding with material in the **interstellar medium**. This produces a shockwave exciting the gas and making it fluoresce.

Hertzsprung-Russell diagram

A graph that plots the brightness of **stars** against their surface temperatures (see pages 52–53). So long as these values are known, then any star can be plotted on the diagram. It was developed by Ejnar Hertzsprung (in Europe) and Henry Norris Russell (in the US), each working independently of the other. From left to right the stars range in color from blue through yellow to red. This corresponds to a decrease in the surface temperature. From top to bottom the stars are ranged from bright to faint so the brightest supergiants are at the top and dim dwarf stars are near the bottom. **Main sequence stars** are represented in a curving line from the top left of the diagram to the bottom right. All stable hydrogen-burning stars, including the **Sun**, are clustered here. The values on the axes are usually chosen so that the Sun is plotted in the center of the diagram. As a star evolves, its position on the diagram moves. Soon after it forms, it joins the main sequence. As the hydrogen in its core is depleted, the star will evolve across the main sequence. As helium begins to fuse in its center, the star becomes a **red giant**. When **fusion** stops, a low-mass star will collapse and become a **white dwarf**, but a high-mass star will explode as a **supernova**. The Hertzsprung-Russell diagram continues to be an important instrument in our understanding of stellar evolution.

High-Speed Photometer (HSP)

An instrument that was originally intended to measure the changes in brightness of celestial objects very rapidly. The HSP was capable of taking a measurement of a celestial object's brightness 50,000 times a second, monitoring changes in an object's luminosity with great precision. Unfortunately, the usefulness of the High Speed Photometer was severely affected by the **spherical aberration** of Hubble's **primary mirror**. It was replaced by the corrective optics unit **COSTAR** during Hubble's first servicing mission.

Hubble constant

A quantity that describes the rate of expansion of the universe per unit of distance. As the universe expands, the objects within it become farther apart. This phenomenon causes **redshift** which can be used to calculate the rate of expansion. Current estimates for this rate vary by a factor of two but lie somewhere between 30–60 miles (50–100 kilometers) per second for every 3.26 million light years. As time goes by, the expansion rate is slowed by the gravity of objects within the universe and the Hubble constant decreases. Measuring a more accurate quantity for the constant is one of the Hubble Space Telescope's primary functions.

Hubble Deep Field

An unprecedented image of a tiny region of sky, showing galaxies in almost every stage of evolution. The image was taken during Christmas 1995 over ten whole days of observing time. The Hubble Deep Field was made available right away to astronomers all around the world so that analysis could be undertaken as quickly as possible. A second Hubble Deep Field, this time of a region of sky south of the Earth, is planned to be imaged over Christmas 1997.

Hubble Tuning Fork

A diagram used to represent the various forms of normal galaxy (see page 78). **Elliptical galaxies** are divided into eight subclasses from E0, which is a sphere, to E7, which is an oblate spheroid (similar to an American football). These subclasses are placed along the handle of the tuning fork. The two prongs of the fork show three subclasses of **spiral galaxy** and three subclasses of **barred-spiral galaxy**. They define the size of the galactic nucleus and also how tightly the arms are wound. Sometimes an intermediate form of galaxy, known as a lenticular galaxy, is placed where the prongs join the handle. No irregular galaxies are shown on the tuning fork.

impact basin

A large crater left by a **meteorite** when it strikes the ground. Impact basins can be seen all over celestial bodies such as the Moon and Mercury because, having no atmospheres, those worlds do not suffer from weathering and erosion. According to theory, the final stages of a planet's accretion are characterized by the bombardment of the protoplanet with **comets** and **asteroids**, producing craters and impact basins. On the Earth, atmospheric erosion and plant life have destroyed the visible signs of most craters.

infrared radiation

Infrared radiation has wavelengths longer than **visible light** and often provides useful information about cool objects within the universe.

interstellar medium

The diffuse material found between the stars within most types of galaxy. The interstellar medium is mainly composed of two elements, dust and gas. However, these elements are not distributed uniformly throughout a galaxy; they tend to gather into **giant molecular clouds**. Within these clouds are **absorption nebulae** and **emission nebulae**. Enhancement in the density of the interstellar medium leads to the production of stars. See **star-forming region**.

irregular galaxy

Any galaxy that does not fit the standard classifications on the **Hubble Tuning Fork** diagram (see page 78) because its shape seems random and disordered. It may have disk-like motion and vestiges of a spiral structure; if so, it is called a type I irregular. Type II irregulars have no hint of structure. Star formation can be seen taking place in some irregular galaxies.

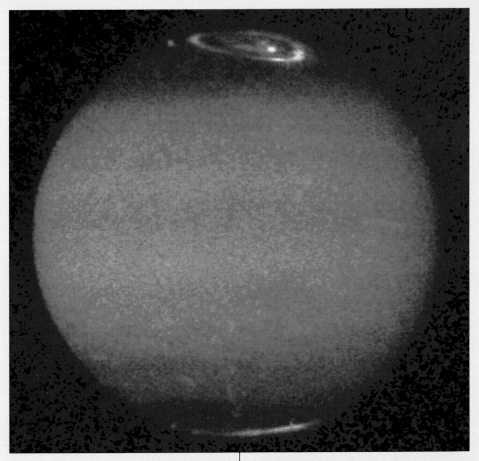

aurora
on the planet Jupiter, taken in ultraviolet light.
The aurora shows up as luminous concentric
rings around each magnetic pole.

emission nebula
in the star-forming region of galaxy NGC 2366.

globular cluster
known as globular cluster G1 in galaxy M31.

near-infrared image
of dwarf star Gliese 105A

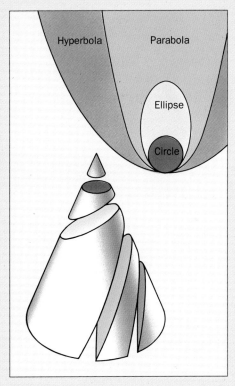

orbit
The orbit of any celestial object is one of the four
shapes; circle, ellipse, parabola or hyperbola.
Collectively they are called conic sections, since
they can all be produced by "slicing" a cone.

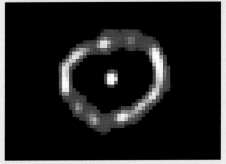

COSTAR
refocuses the light from Hubble's primary mirror,
correcting the spherical aberration. The image
(top) was taken before COSTAR; the second
image (above) shows the same object, Nova
Cygni, photographed after COSTAR was installed.

115

light year

The distance that light travels in one year. It is equivalent to 5.87 thousand billion miles (9.46 thousand billion kilometers). Distances in the universe are so large that light years are frequently the most convenient measurement. For example, the nearest star to the **Sun** is Alpha Centauri, over four light years away. The **Milky Way** is 100,000 light years across and the diameters of clusters of galaxies are measured in millions of light years.

Local Group

The very small cluster of galaxies containing a few dozen members, of which our own galaxy is one. There are three spiral galaxies: the Milky Way, Andromeda and M33. A giant elliptical galaxy, Maffei I, and small ellipticals and irregular galaxies are also members. A barred-spiral galaxy, Dwingaloo 1, has been discovered to lie on the very outskirts of the group. The shape of the Local Group is like a dog's bone with concentrations of galaxies around the **Milky Way** and Andromeda. Its radius is approximately 3.25 million light years. It is a small part of a much larger conglomeration known as the Local Supercluster which contains, among others, the Virgo cluster of galaxies.

look-back time

The time taken for light to reach the Earth from a distant celestial object. This also indicates how far into the past we are looking when we see an image of the object. For example, when light reaches us from an object that is 20 million **light years** away, the image it produces shows us the object as it looked, 20 million years ago. Almost certainly, the way it really looks today is nothing like that image, but it will take another 20 million years before we can find out.

low Earth orbit

Orbit at an altitude of 320 miles (525 kilometers), in which an object takes approximately 90 minutes to circle the Earth. The Hubble Space Telescope is in low Earth orbit, chosen so that the telescope is within easy reach of the **Space Shuttle**, which operates only in low Earth orbit. Accessibility by the Space Shuttle is necessary because Hubble's instruments

need to be replaced or repaired. Periodically, the telescope also needs boosting into a slightly higher orbit, to counteract its gradual loss of height and prevent it from burning up in Earth's atmosphere.

main sequence star

Any star in its stable, "middle-aged" period of existence, usually defined by astronomers as a star that is fusing hydrogen at its core. The main sequence classification refers to the fact that when these stars are plotted on a **Hertzsprung-Russell diagram** (see pages 52–53), they form a curving "S"-shaped band that runs diagonally right across the diagram. When hydrogen ignites in a **protostar**, the star passes through an unstable phase before arriving at the main sequence. As the star ages and uses up its hydrogen, its luminosity changes and it moves across the main sequence. As the hydrogen burning in its core runs out, the star will leave the main sequence and become a **red giant**.

mass

A fundamental property of matter. The amount of mass contained in an object and its density result in the strength of its gravitational field.

meteor

A particle of matter or a small pebble, which approaches a planet and burns up in the atmosphere of that world. When observed from Earth it is commonly referred to as a shooting star. There are some very well known meteor showers that collide with Earth's atmosphere every year. Most have had their origins traced to short-period **comets** that litter their orbits with particles from their dust tails. Other meteors are thought to be produced by collisions between **asteroids**.

meteorite

Any solid object that enters a planet's atmosphere and instead of burning up, actually strikes the planet's surface.

meteoroid

Any particle of dust, or larger particle, found in the **Solar System.** Periodically, meteoroids enter a planet's atmosphere and become **meteors** or **meteorites.**

Milky Way

The name of our own **galaxy**, which contains about 100 million stars. Analysis has shown that it is certainly not an **elliptical galaxy**. There is still controversy over whether it is a **spiral** or a **barred-spiral galaxy**. At present, the **Sun** exists in the trailing edge of the Orion spiral arm about two-thirds of the way toward the edge of the Milky Way. All the stars visible in the night sky are in our own galaxy. The misty band of light, which stretches across the night sky and is traditionally called the Milky Way, is our view of the spiral arms. The Milky Way has a diameter of 100,000 **light years**. It possesses a halo of **globular clusters** and is held by gravity to a few dozen other galaxies, known as the **Local Group**.

minor bodies

Any object in the **Solar System** that is neither a **planet** nor a **meteoroid**. Minor bodies are usually **asteroids** or **comets**.

moon

Any large, naturally occurring object which is in **orbit** around a **planet**. Most moons are tiny compared to their parent planets. Earth's Moon and Pluto's moon, Charon, are relatively large moons.

NASA

The National Aeronautics and Space Administration, a US government-funded body in charge of the national space exploration program. It is part of NASA's responsibility to oversee the **Space Shuttle** and Hubble Space Telescope programs. It is also responsible for sending **probes** to explore other planets in our **Solar System**. NASA supports a number of astronomy projects and funds related research into new space technology.

Near-Infrared Camera and Multi-Object Spectrometer (NICMOS)

A new instrument installed on the Hubble Space Telescope during the second servicing mission, in February 1997. The Near-Infrared Camera and Multi-Object Spectrometer should be capable of supplying astronomers with images that are far in advance of any currently available today. With this camera, astronomers will be able to explore the dustiest regions of the universe and search for stars and other bodies in the process of formation. The dusty central regions of **active galactic nuclei** will also be easier to explore with this instrument.

near-infrared image

An image taken with a camera designed to detect wavelengths just beyond those of visible red light. Near-infrared images are particularly useful for showing objects and environments that are too cool (typically less than 1,000°C) to be detectable in **visible light**. They are also used to reveal objects that are obscured by surrounding regions of dust. *See image on page 115.*

nebula

A cloud of dust and gas in space that is visible to observers on Earth because it emits, reflects or absorbs starlight. **Emission nebulae** can be created by **star-forming regions** or the processes that govern stellar death. When high-mass stars explode as **supernovae**, they create emission nebulae known as supernova remnants. When low-mass stars die, they create emission nebulae that are known as **planetary nebulae** even though they have nothing to do with planets. Reflection nebulae are clouds of dust that are not illuminated from within, but scatter the light from nearby stars in our direction. **Absorption nebulae** are areas of denser-than-average material that obscure background light.

Neutral Buoyancy Simulator

A large tank of water used by astronauts in training exercises when they rehearse maneuvers to be completed in space. A neutral buoyancy simulator is the best way to simulate the weightless conditions of space for extended periods of time. Although trainee astronauts still feel the downward pull of **gravity**, it is somewhat compensated for by the buoyancy they feel from the water.

neutron star

The central remains of a star following a **supernova** explosion. A neutron star is composed of very high density matter that has undergone a dramatic gravitational collapse. Typically a neutron star has a radius of just 6.2 miles (10 km), but a density of up to 10^{18} kg/m^3. They were once thought to be impossible to detect because of their small size and low energy. See **pulsar**.

observatory

A building or a collection of buildings housing a telescope and its ancillary equipment. The best observatories in the world are in very high locations (on a mountain top or on the Antarctic Plateau) which means that they are above most of the distorting effects of the atmosphere. It also puts them above the cloudbase and weather systems, which plague ground-based astronomy.

open cluster

A cluster of stars all born from the same cloud of collapsing dust at approximately the same time. As these open clusters orbit the center of the **galaxy**, the individual members split up and move farther apart. Open clusters contain relatively young stars which are (at most) a few thousand million years old.

orbit

The path of a celestial body or **satellite** under the influence of **gravity**. The shape of an orbit is always one of the four mathematical shapes known as conic sections. This generic term for circles, ellipses, parabolae and hyperbolae derives from the fact that all these shapes can be produced by taking a cone and "slicing" it in various ways. The shapes can be referenced by a value known as **eccentricity**. A circle is the least eccentric shape and a hyperbola is the most eccentric shape. *See diagram on page 115.*

parallax

The apparent angular displacement of an object, caused by viewing it from slightly different positions. Parallax is used in astronomy to determine the distance to nearby stars. A star is observed from Earth on two occasions, six months apart. These two readings provide the maximum possible difference in position from opposite points in the diameter of the Earth's orbit. The angular displacement of the star can then be used to calculate its distance using trigonometry.

peculiar galaxy

Any **galaxy** that is strange or distorted in a way that causes it to deviate from the pattern of normal galaxies. Most peculiar galaxies are **starburst** galaxies or galaxies that are being affected by the gravity of another galaxy close by. Some **radio galaxies** also appear to be distorted.

pixel

A picture element on a **charge-coupled device** (CCD) or computer monitor. It is the smallest piece of picture information stored on a computer. An image is made up of thousands of pixels, each of which contains information regarding that particular point's brightness. On computer screens, pixels show color by varying the brightness of their red, green and blue components.

planet

A rocky, gaseous or icy body that is in orbit round our Sun or around any other star. In our own **Solar System** there are nine planets: Mercury, Venus, Earth, Mars, Jupiter, Saturn, Uranus, Neptune and Pluto. *See image on page 121.*

planetary nebula

The outer, gaseous layers of a **red giant**, which have been blown off into space while the stellar core is in the process of becoming a **white dwarf**. The astronomer William Herschel called these layers planetary nebulae because of their resemblance to **planets**. They are constantly expanding and have estimated lifespans of less than 40,000 years. After this time they become too tenuous to be seen. The reason they shine is because the gas within them is excited by electromagnetic radiation from the central, collapsing star. See **nebula**.

primary mirror

The first mirror encountered by light rays in a reflecting telescope. The primary mirror is also the larger mirror in a two-mirror telescope system. It is usually parabolic in shape and focuses light to a point near the top of the telescope's tube. Here it is either collected directly or reflected by another mirror (known as the **secondary mirror**) to a more convenient location for collection.

probe
A robotic spacecraft designed to travel to places that are impractical or inaccessible for humans to reach themselves. The highly successful Pioneer and **Voyager probes** were sent by **NASA** to survey other planets in our Solar System, most notably the outer gas giants. Other high-profile probes have been the Russian Venera spacecraft, which landed on the surface of Venus, and NASA's Viking probes, which landed on the surface of Mars. Current NASA probes include **Galileo**, which is on a mission to Jupiter.

proplyd
An abbreviation of the term protoplanetary disk. A proplyd is the dusty **accretion disk** around a **protostar**, inside which planets are thought to be forming. Examples of proplyds were found by the Hubble Space Telescope in the Orion nebula, which is the region of on-going star formation closest to Earth.

protoplanetary disk
See proplyd.

protostar
The name given to a star in the process of formation. Astronomers make detailed observations of these objects on a continuous basis. Protostars are found in the dense areas of **giant molecular clouds**. They are still involved in the accretion process, and nuclear reactions have not yet begun within their cores. As they are surrounded by large "envelopes" of dusty matter, they must be studied with long wavelengths of electromagnetic radiation, such as infrared.

pulsar
A spinning **neutron star**. A pulsar is made visible to us because a beam of electromagnetic radiation emanates from each magnetic pole. The magnetic axis of the pulsar is not aligned with its rotation axis and so as the pulsar rotates, it emits a beam of radiation, tracing an oval shape across the night sky. The Earth lies at one point on that oval, and so we see the pulsar's light shining on and off, similar to the way that a lighthouse's rotating light appears to flash at shipping. There are two types of pulsars. One type spins

with a relatively slow rotation rate, for example, the pulsar in the Crab nebula supernova remnant. This type slows down as time passes. The second type is found in binary star systems and spins progressively faster as material is transferred from the companion star. This type is known as a millisecond pulsar. Both pulsar types are created from the remains of a star that has exploded as a **supernova**.

quasar
A contraction of "quasi-stellar object", a compact and very distant star-like object which is highly luminous, and an excellent source of radio waves. A quasar is just one of the many types of **active galaxies** that astronomers have identified. Early detailed analysis suggested that they were **active galactic nuclei**, even though the surrounding galaxies could not be seen because of the blinding effect of the quasar's light. (Some are 100 times more luminous than the brightest galaxy). Some of the most recent observations from the Hubble Space Telescope have finally shown what the galaxies that surround quasars look like. A few quasars appear to reside in very distant normal spiral and elliptical galaxies, but the vast majority of quasar hosts are **colliding galaxies**. *See images on page 121.*

radio galaxy
A galaxy that is an intense source of radio waves. Radio galaxies are a type of **active galaxy** and are invariably elliptical in shape. The radio emission usually comes from two huge lobes on either side of the galaxy. The lobes themselves are vast, and can measure millions of light years across. Radio waves are produced when energetic particles shoot out from the **active galactic nucleus** and spiral around magnetic field lines.

radio telescope
A telescope specifically designed to observe radio waves. A radio telescope is a large dish-like antenna that focuses radio waves toward a detector, mounted in front of the dish. Two or more radio telescopes can be used in conjunction to form an interferometer, which increases the amount of detail that can be resolved. See **radio galaxy**.

red dwarf
Any star on the **Hertzsprung-Russell diagram** that has a low surface temperature and is found on the lower right of the **main sequence**. Red dwarfs are thought to be more numerous in the universe than any other type of star. It is not easy to check this theory, however, because they are so small and dim that they are notoriously difficult to detect.

red giant
Any star that is burning helium in its core. The onset of helium **fusion** causes a star to leave the **main sequence** and become a red giant. These red giant stars are found in the upper right corner of the **Hertzsprung-Russell diagram**. Although their red color is an indication of relatively low surface temperature, these stars are so large that the total amount of energy that they radiate is enormous. Typically, a red giant's radius is about the radius of Mars' orbit.

redshift
The lengthening of the wavelength of light from a star or galaxy caused, either by the motion of the source away from the observer, or the motion of the observer away from the source. When the light from distant galaxies is redshifted, light in the visible light spectrum appears more red as a result (see diagram on page 92). This is interpreted to mean that the galaxies are moving away from us, due to the expansion of the universe. Edwin Hubble showed that the farther away a celestial object is, the greater the amount of redshift it registers. This fact can be used to help measure the distance to faraway galaxies. See **Doppler effect**.

satellite
Any object in **orbit** around a **planet**. A satellite can either be natural (a moon), or artificial (a spacecraft). Several of Saturn's moons are known as shepherd satellites, because they keep some components of Saturn's rings in orbit by the force of their gravity.

secondary mirror
The smaller of the two mirrors usually found in a reflecting telescope. The secondary mirror is responsible for reflecting

the light back down the tube of the telescope to a focus at the bottom, near the **primary mirror**. Usually, the primary mirror has a small hole drilled in its center to let the focused light from the secondary mirror pass through.

Seyfert galaxy

A specific type of **spiral** or **barred-spiral galaxy**, which has an **active galactic nucleus**. The nucleus far outshines the spiral arms of the host galaxy, and the electromagnetic radiation released by a Seyfert nucleus is not in the form of starlight. These observations have led to the idea that a Seyfert Galaxy is a type of **active galaxy**. Analysis has shown that the nuclei of Seyfert galaxies contain hydrogen clouds that are swirling around at very high velocities. Certain types of Seyferts resemble **quasars** in many ways, but are slightly less powerful and much closer to us.

solar panels

Devices that transform the light from the Sun into electrical power. The Hubble Space Telescope possesses two arrays of solar panels to provide all the power needed for the telescope to perform its observations and to send the data it has collected to Earth.

Solar System

The group of celestial bodies governed by the Sun's gravitational field. The Solar System is made up of the **Sun**, the nine **planets** and their **moons**, and a number of **minor bodies** such as **asteroids** and **comets**. Experts believe that it was formed from a collapsing **absorption nebula** some 4.5 billion years ago. It is often considered as two separate regions: the inner Solar System, consisting of the four rocky planets (Mercury, Venus, Earth and Mars); and the outer Solar System, containing the gas giants (Jupiter, Saturn, Uranus and Neptune) as well as the outermost planet – small, icy Pluto.

Space Shuttle

NASA's reusable spacecraft used to transport objects and people into **low Earth orbit**. Six Space Shuttles have been constructed. The first was called Enterprise (named after the starship in the television series Star Trek) and was a prototype that never flew in space. Enterprise was followed by Columbia, Challenger, Discovery, Endeavour and Atlantis, all named after famous sailing ships. Challenger was lost in a tragic launch accident in January 1986 that resulted in the death of seven astronauts.

Space Telescope Imaging Spectrograph (STIS)

A new spectrograph installed in the instrument cluster on board the Hubble Space Telescope during the second servicing mission in February 1997. The Space Telescope Imaging Spectrograph can span **ultraviolet**, **visible-light** and **near-infrared** wavelengths, and should provide astronomers with a wealth of new information about celestial objects.

spectrograph

A device that splits light from a celestial object into a spectrum, allowing its chemical composition to be studied. Spectrographs are sometimes referred to as spectrometers. Originally there were two spectrographs on board the Hubble Space Telescope. A new spectrograph was installed in place of old instrumentation during the second servicing mission of February 1997.

spherical aberration

A flawed condition of curved mirrors, most notably spherical mirrors, in which the light reflected from the outer edges comes to a focus at a different point from the light reflected by the central regions of the mirror. The **primary mirror** of the Hubble Space Telescope suffers from a spherical aberration due to an error when grinding the mirror.

spiral galaxy

Any **galaxy** in which a central bulge of older stars is surrounded by a flattened galactic disk containing young, hot stars in a spiral pattern. Spiral galaxies are closely related to **barred-spiral galaxies** (see page 78).

star

Any celestial body generating energy through the process of nuclear **fusion** taking place in its core. The generic term star is also used to describe the **stellar remnant** known as a **white dwarf**. Stars exist in a variety of sizes, characterized by the mass they contain. A truly massive star may contain between 50 and 100 times the mass of the **Sun**. These **blue supergiants** burn furiously and only last for a few million years. The least massive stars are **red dwarfs**, which contain only a few tenths of the mass of the Sun. They burn much more slowly and live for many billions of years. Every type of star can be plotted on the **Hertzsprung-Russell diagram** (see pages 52–53).

starburst

A massive bout of star formation occurring within a galaxy. It can be triggered when **giant molecular clouds** collide with each other during galaxy mergers. Electromagnetic radiation is emitted copiously at many wavelengths, including ultraviolet radiation from **blue supergiants** and infrared radiation from dust, which has been heated by the starburst. See **star-forming region**.

star-forming region

Any dense region of **interstellar medium** where stars are actively forming. Star-forming regions can usually be located using **emission nebulae** as indicators. Younger star-forming regions are found in **absorption nebulae**. In **spiral galaxies**, the principal sites of star formation are on the leading edges of the spiral arms. Low-mass stars will be more numerous in this region, but the easiest stars to see will be the brighter, but rarer, high-mass stars. See **nebula**.

stellar remnant

The collapsed remains of a dead star. The precise properties of a stellar remnant depend on the initial mass of the dying star. A low-mass star produces a **white dwarf** with a mass of about 1.5 times that of the Sun. The collapse of a high-mass star produces a stellar remnant that in its turn becomes either a **neutron star** or a **black hole**. A less massive remnant produces a neutron star with a mass of 1.5–3 times that of the Sun; a more massive remnant (greater than 3 times the mass of the Sun) produces a black hole. See **supernova**.

stellar wind

A stream of subatomic particles that flows away from a star along magnetic field lines. A stellar wind is composed of charged particles such as protons, electrons and the atomic nuclei of helium.

Sun

The star at the center of the **Solar System**, around which the Earth and other planets orbit. The Sun is a **yellow dwarf** star with a diameter that is over 100 times that of the Earth. Experts believe that it was formed by accretion in the gravitational collapse of an **absorption nebula**. Products of radioactive decay in **asteroids** indicate that the gravitational collapse was possibly triggered by a nearby **supernova**, some 4.5 billion years ago.

supercluster

A conglomeration of clusters of galaxies that is held together by the force of their mutual gravity. Superclusters stretch for hundreds of millions of light years through the universe. So far about 50 separate superclusters have been identified by astronomers.

supernova

A catastrophic explosion that blows a **star** to pieces. There are two types of supernova. Supernovae type I are thought to begin as **white dwarfs** in binary star systems. They slowly accrete material until a catastrophic nuclear explosion blows the star to pieces. Type II supernovae occur in stars that have more than five times the mass of the Sun. The nuclear **fusion** in the core of such a star produces all the elements on the periodic table up to and including iron; then the fusion process can go no further. Iron is so stable that it requires a huge input of energy before it can fuse. The core is inert for a period of time, but the iron continues to accumulate. When the iron in the core reaches 1.4 times the mass of the Sun (known as the Chandrasekhar limit), gravity overwhelms it and the star collapses. This provides the energy to explode the star and synthesize the elements beyond iron on the periodic table. Type I supernovae are even more luminous than type II, but both types are so bright that they outshine the galaxy in which they are located. A supernova will fade over a period of months, as the original energy of the explosion is replaced by the radioactive decay of unstable atomic nuclei. If it is not disrupted altogether, the central region becomes either a **neutron star** or a **black hole**. The supernova's outer layers are blown off into space to form a **nebula**, which is often referred to as a supernova remnant.

torus

The large doughnut-shaped concentration of dust and gas that surrounds the **black hole** and **accretion disk** in a typical **active galactic nucleus**. The torus exists in the equatorial plane of the black hole and (in certain types of active galaxy) prevents astronomers from seeing this region. Although the Hubble Space Telescope has taken the first ever image of a torus (in NGC 4261), astronomers are still unsure of how and why a torus forms.

Tracking and Data Relay Satellite System (TDRSS)

A set of communications satellites in **geostationary orbit** above specific locations on the Earth.

true-color image

Any astronomical image in which the colors closely resemble what would be seen with the naked eye (see **false-color image**). Three color plates are combined to produce a true-color image; red, blue and green. *See image on page 121.*

ultraviolet radiation

A band of electromagnetic wavelengths that are shorter than the blue end of the **visible-light** spectrum. Ultraviolet radiation is quite energetic and is given out in copious quantities by **blue supergiant** stars. **Emission nebulae** around star-forming regions usually shine because their constituent gas is being excited by ultraviolet radiation. Some of Hubble's instruments work in the ultraviolet spectrum and can produce images to show what objects look like in ultraviolet light.

visible light

The band of electromagnetic wavelengths to which our eyes are sensitive. Visible light wavelengths run from violet to red with the different wavelengths of radiation appearing to us as different colors. Ultraviolet light is off the top of the visible light scale and infrared radiation is off the bottom of the scale.

Voyager probes

Two unmanned **NASA** space probes, both launched in 1977, that traveled to the gas giants of the outer **Solar System**. Both sent back detailed information and images of Jupiter, Saturn and their moons. Voyager 2 went on to provide the first ever close-up photographs of Uranus (1986) and Neptune (1989).

white dwarf

A highly evolved **stellar remnant**, part of the remains of a star after the process of nuclear **fusion** has finished. White dwarfs are often found in **planetary nebulae** and range in surface temperatures from 4,000 to 15,000 K. They have no internal heat source, but radiate their residual energy until they become black dwarfs. They are very small, some having a diameter less than that of the Earth. A white dwarf in a binary system could produce a type I **supernova**.

Wide Field and Planetary Camera (WFPC)

The principal imaging instrument on the Hubble Space Telescope. The Wide Field and Planetary Camera is not housed in the telescope's instrument cluster; instead, it is located in the side of the telescope. It contains four individual **charge-coupled devices** which work in tandem to give a wide field of view to study extended objects. One of the CCDs is operated at a higher magnification and this is known as the planetary camera. Its images can be either incorporated with those of the three wide field cameras or used in isolation. The Wide Field and Planetary Camera was completely replaced during the first servicing mission. The new version contains optics to correct the Space Telescope's **spherical aberration**. *See images on page 121.*

yellow dwarf

Any star on the **main sequence** that has a surface temperature of approximately 6,000 K. The **Sun** is a yellow dwarf.

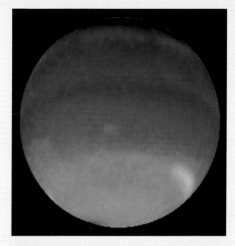

planet
Neptune in false colors, released by NASA in October 1996. High banks of cloud appear white, and the area near the equator, where winds blow at 900 mph (1,448 km/h), appears dark blue.

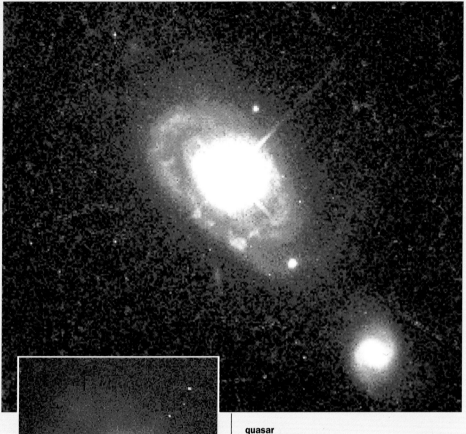

quasar
PG 0052-251 (above), at the core of a spiral galaxy, some 1.4 billion light years from Earth. Quasar IRAS13218+0522 (left) is two billion light years away, where two galaxies are merging. They appear to have orbited each other before colliding, leaving loops of glowing gas.

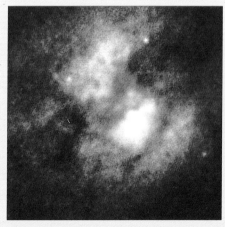

Wide Field and Planetary Camera (WFPC)
in operation before and after the first servicing mission. The image (top) shows the very bright star, Melnick 34, in the 30 Doradus region, shot with the original camera. The second image (above) of the same star, demonstrates how much more focused WFPC2 is by comparison.

barred-spiral galaxy
NGC 4639, in the Virgo cluster of galaxies, some 78 million light years from Earth.

true-color image
of the starburst galaxy Arp 220, showing gigantic young star clusters at its core.

Astronaut Steven L. Smith accessing Hubble's instrument cluster during the second servicing mission which took place during February 11–21, 1997. The Telescope's damaged thermal covering *(inset)* was also repaired during the mission.

INDEX

FURTHER READING

Booth, Nicholas *Exploring the Solar System* (Cambridge University Press, 1996)

Cattermole, Peter *Earth and Other Planets* (Andromeda, 1995)

Chapman, R. and Brandt, J. *The Comet Book* (Jones and Bartlett, 1984)

Chown, Marcus *Afterglow of Creation* (Arrow Books, London, 1993)

Clark, Stuart *Redshift* (University of Hertfordshire Press, 1997)

Clark, Stuart *Towards the Edge of the Universe* (Wiley-Praxis, 1997)

Clark, Stuart *Stars and Atoms* (Andromeda, 1995)

Ferguson, Kitty *Prisons of Light* (Cambridge University Press, 1996)

Ferris, Timothy *Galaxies* (Bantam Press, 1988)

Goodwin, Simon *Hubble's Universe* (US ed. Viking, 1996; UK ed. Constable, 1996)

Kauffman, William J. III *Universe* (W.H. Freeman and Co., Oxford, 1985)

Kippenhahn, R. *100 Billion Suns* (Weidenfeld and Nicholson, London, 1983)

Kitchin, C.R. *Journeys to the Ends of the Universe* (Adam Hilger, 1990)

Leverington, David *A History of Astronomy* (Springer, 1996)

Levy, David *The Sky: A User's Guide* (Cambridge University Press, 1991)

Littman, Mark *Planets Beyond* (Wiley, 1988)

Mitton, Jacqueline *Dictionary of Astronomy* (Penguin, 1993)

Moore, Patrick *Atlas of the Universe* (Philips, 1994)

Moore, Patrick *New Guide to the Planets* (Sidgwick and Jackson, 1993)

Moore, Patrick (ed) *The Astronomy Encyclopedia* (Mitchell Beazley, 1987)

Nicolson, Iain and Moore, Patrick *The Universe* (Collins, London, 1985)

Pasachoff, J.M., Spinrad, H., Osmer, Patrick S. and Cheng E. *The Farthest Things in the Universe* (Cambridge University Press,1994)

Petersen, Carolyn Collins and Brandt, John C. *Hubble Vision* (Cambridge University Press, 1995)

Vanin, Gabriele *A Photographic Tour of the Universe* (Firefly, 1996)

Weinberg, Steven *The First Three Minutes* (Andre Deutsch, London, 1977)

Zukav, Gary *The Dancing Wu Li Masters* (Flamingo, 1979)

ACKNOWLEDGMENTS

Abbreviations: **NASA** National Aeronautics and Space Administration, **SPL** Science Photo Library

All photo credits NASA except:
12tr The Huntington Library, San Marino, California
24 NASA/Spacecharts; **25br** Spacecharts; **31t** NASA/SPL; **57r** National Optical Astronomical Observatories/SPL; **84bl** Space Telescope Science Institute/NASA/SPL;
98t Space Telescope Science Institute/NASA/SPL;
108 Roger Ressmeyer/Starlight/SPL; **109** European Southern Observatory; **122** NASA/SPL
Jacket: *Front background image* Space Telescope Science Institute/NASA/SPL

Artists: Julian Baker, Ron Brocklehurst, Steve McCurdy
Picture research: Diane Hamilton